BARRON'S

# PAINLESS
# Chemistry

Loris Chen

# Dedication

This book is dedicated to my children Jay and Tiffany and the students who asked not "What?" but rather "How?" and "Why?"

*All inquiries should be addressed to:*
Barron's Educational Series, Inc.
250 Wireless Boulevard
Hauppauge, New York 11788
**www.barronseduc.com**

ISBN: 978-0-7641-4602-2

Library of Congress Control Number: 2011005822

**Library of Congress Cataloging-in-Publication Data**

Chen, Loris
    Painless physical sciences chemistry / Loris Chen.
       p. cm. — (Painless series)
    Includes index.
    ISBN: 978-0-7641-4602-2 (pbk.)
  1. Chemistry—Study and teaching (Middle school).    I. Title.

QD40.C46    2011
540—dc22

                                                      2011005822

PRINTED IN THE UNITED STATES OF AMERICA
9 8 7 6 5 4 3 2

# CONTENTS

# Chapter 5: Mixtures 95

# Chapter 6: Molecules and Compounds 133

# Chapter 7: Chemical Reactions 163

# Chapter 8: Acids, Bases, and Salts 191

# INTRODUCTION

Cake is chemistry. Flour, sugar, baking soda, baking powder, water, oil, and eggs are all made of molecules composed of atoms. The combinations of atoms give each ingredient its unique characteristics. Molecular arrangement explains why sugar is sweet but flour that is made of many sugar units is not. Chemistry explains why baking soda is bitter but water is not, and why oil and water won't mix without some help from the other substances.

When flour, sugar, baking soda, and baking powder are sifted together, they form a mixture of solids. Although mixed, each substance keeps its original characteristics. It might take some time, but they can be separated from each other. Add water and oil, and the mixture becomes an emulsion. Add eggs and a slow chemical reaction begins to form tiny air bubbles.

To speed up the chemical reaction, bake the batter at 350°F for 30 minutes. If you live at high altitude, you will need to adjust the baking temperature and time to compensate for reduced air pressure—more chemistry.

The formation of gas bubbles and the changes in aroma, color, and texture are signs that a chemical reaction has taken place. The ingredients can no longer be separated easily and no longer have their individual characteristics.

As you eat the cake, a series of chemical reactions digests the particles into the building blocks of life needed for protein synthesis and respiration. From the food you eat to the air you breathe, chemistry is everywhere.

Painless Chemistry is organized around the big ideas of chemistry. Each chapter contains examples, memory hints, and caution notes. Whenever possible, diagrams and graphs are used to explain some of the harder concepts. At the end of each section, you will find Brain Ticklers that test your understanding the big ideas.

During middle school you may not use all the information contained in Painless Chemistry. Each chapter of the book includes basic information for now and more advanced information for later. If the math is too difficult, focus on the concepts instead. Don't worry about memorizing the periodic table of the elements. It's a tool. Learn instead how to use it. Most of all remember, that with effort, what seems difficult now will seem easy later.

# CHAPTER ONE

# Matter

# ENERGY OR MATTER

**Chemistry** is the division of physical science that studies the composition, properties, and reactions of substances. The nature of matter and energy have been investigated and debated from ancient times.

Everything is either energy or matter. **Energy** is the ability for matter to do work. Energy can affect the rate of a chemical reaction. Chemical reactions can convert one type of energy into another type. Some examples of energy associated with chemical reactions are light, heat, and electricity.

**Matter** is defined by the physical characteristics of mass and volume. In other words, matter has substance and takes up space. The amount of matter in a specified space is density. For pure substances, density is an identifying characteristic.

# MASS

**Mass** is the amount of matter in an object. The basic unit of mass is the **kilogram**. One gram is the approximate mass of a 5-cm (2-inch) steel paper clip. A kilogram is 1,000 grams (2.2 pounds). One liter of water has a mass of 1 kilogram.

Common tools for measuring mass are the double pan balance, triple beam balance, and electronic balance. The double pan balance compares the mass of a substance to a known standard mass (a metric weight). The triple beam balance has an arm with sliding mass units that are adjusted to balance the mass of a substance. The electronic balance provides a digital measurement of the mass of a substance.

## Mass Versus Weight

**Weight** is a measure of the pull or acceleration of gravity on an object's mass (w = mg, or weight equals mass times the acceleration of gravity). The acceleration of gravity near Earth's surface is $9.8 \text{ m/sec}^2$. The weight of an object can change if gravity changes. The amount of matter in an object does not change even if the pull of gravity changes. Therefore the mass of an object does not change even if the pull of gravity changes.

In many countries, the kilogram (kg) is commonly used to express weight. In science, kilogram describes mass. The newton is a measure of metric weight and is the force needed to accelerate 1 kilogram of mass at a rate of 1 meter per second per second. The unit of measure of a newton is $kg \cdot m/sec^2$. One pound is about 4 newtons.

Sometimes in chemistry the term *weight* is used when *mass* is meant. Be careful to pay attention to units of measure.

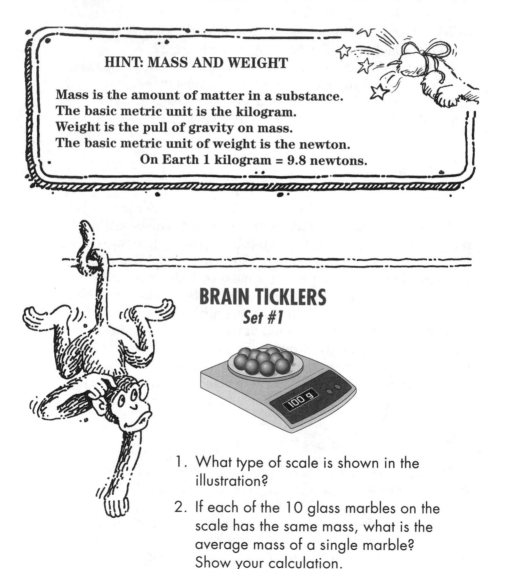

## HINT: MASS AND WEIGHT

**Mass is the amount of matter in a substance.**
**The basic metric unit is the kilogram.**
**Weight is the pull of gravity on mass.**
**The basic metric unit of weight is the newton.**
**On Earth 1 kilogram = 9.8 newtons.**

# BRAIN TICKLERS
## Set #1

1. What type of scale is shown in the illustration?

2. If each of the 10 glass marbles on the scale has the same mass, what is the average mass of a single marble? Show your calculation.

3. A brass plate has a mass of 2 kilograms on Earth. The gravity of the moon is one-sixth that of Earth's gravity. What is the mass of the brass plate on the moon? Explain your answer.

(Answers are on page 19.)

# VOLUME

**Volume** is the space that matter occupies or takes up. Volume is measured in three dimensions: length, width, and height. A common tool for measuring the volume of a regularly shaped solid is the metric ruler or **meter stick**. Common units of measure for solids are cubic centimeters ($cm^3$) and cubic meters ($m^3$).

A common tool for measuring liquid volume is the **graduated cylinder**. The volume is measured at the bottom of the **meniscus**, or curve of the liquid in the cylinder. Common measures of liquid volume are milliliter (ml) and liter (l). One milliliter is equal to 1 cubic centimeter ($cm^3$). A liter is 0.26 gallons, or a little more than a quart.

**Tools for measuring volume**

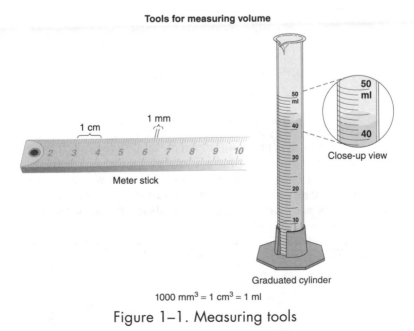

Meter stick

Close-up view

Graduated cylinder

$1000\ mm^3 = 1\ cm^3 = 1\ ml$

Figure 1–1. Measuring tools

# HINT: VOLUME

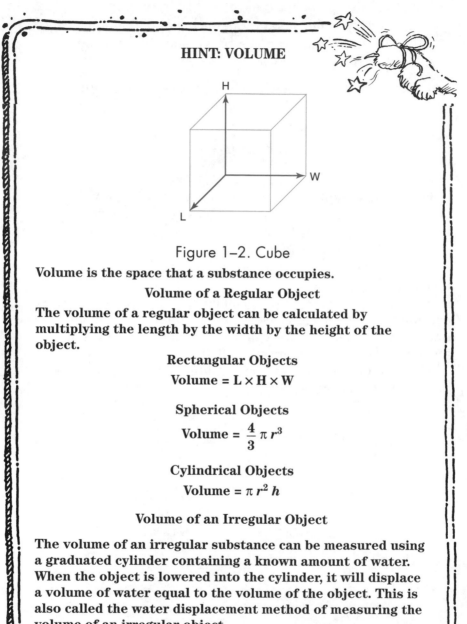

Figure 1–2. Cube

Volume is the space that a substance occupies.

## Volume of a Regular Object

The volume of a regular object can be calculated by multiplying the length by the width by the height of the object.

### Rectangular Objects
Volume = $L \times H \times W$

### Spherical Objects
Volume = $\frac{4}{3} \pi r^3$

### Cylindrical Objects
Volume = $\pi r^2 h$

## Volume of an Irregular Object

The volume of an irregular substance can be measured using a graduated cylinder containing a known amount of water. When the object is lowered into the cylinder, it will displace a volume of water equal to the volume of the object. This is also called the water displacement method of measuring the volume of an irregular object

# BRAIN TICKLERS
### *Set #2*

1. An irregularly shaped chunk of granite
   was lowered into a graduated cylinder
   containing 30 milliliters of water. The
   volume increased to 34 milliliters. What
   is the volume of the granite chunk?

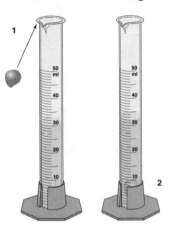

2. What is the volume of the box? Include the unit of measure
   in your answer.

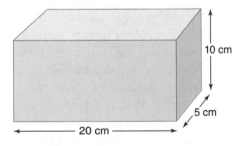

3. What tool would you use to measure the length, width,
   and height of the box?

4. The radius of a metal ball is 2 cm. What is the volume of
   the ball? Describe two ways to find the volume of the ball.

(Answers are on page 19.)

# DENSITY

The amount of matter in a given space is **density**. Density is a calculated physical property of matter that is derived from measurements of mass and volume. Density is mass divided by volume.

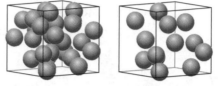

Substance A          Substance B

Figure 1–3. Density diagram

Substance A and substance B have the same volume, 1 cm$^3$. Substance A has a mass of 24 grams. Substance B has a mass of 12 grams. The density of substance A is 24 g/cm$^3$ and the density of substance B is 12 g/cm$^3$.

$$\text{Density substance A} = \frac{24\,g}{1\,cm^3} \qquad \text{Density substance B} = \frac{12\,g}{1\,cm^3}$$

Figure 1–4. Density of A and B

Substance A is denser than substance B. In other words, there is more matter in the same amount of space.

Solids and liquids cannot be compressed. A sample of a solid or liquid has a definite mass and volume. Because density is a ratio of mass to volume, the density of a solid or liquid substance is constant regardless of the size of a sample.

Density can be used to identify a pure solid or liquid substance. All samples of substance A will have a density of 24 g/cm$^3$ regardless of the volume of the sample. All samples of substance B will have a density of 12 g/cm$^3$ regardless of the volume of the sample.

Gases can be compressed. A sample of a gas has a definite mass, but the volume can vary, depending on temperature and pressure. For density to be an identifying property of a gas, the pressure and temperature need to be specified.

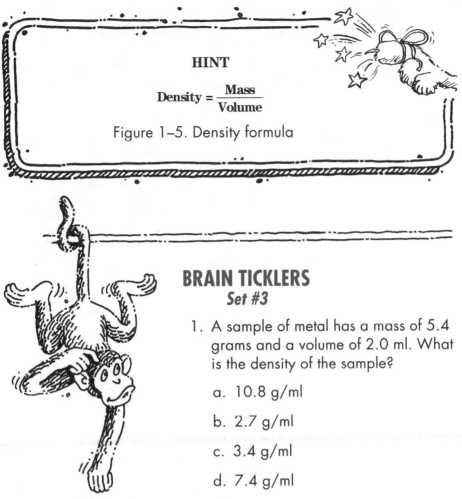

## HINT

$$\text{Density} = \frac{\text{Mass}}{\text{Volume}}$$

Figure 1–5. Density formula

## BRAIN TICKLERS
### Set #3

1. A sample of metal has a mass of 5.4 grams and a volume of 2.0 ml. What is the density of the sample?

   a. 10.8 g/ml

   b. 2.7 g/ml

   c. 3.4 g/ml

   d. 7.4 g/ml

2. If the density of a post-1982 penny is approximately 6.94 g/ml and the mass is 2.5 g, what is the volume of a penny?

3. The density of pure copper is 8.96 g/cm³. Is a penny made of pure copper? Explain your answer.

(Answers are on page 20.)

# METRIC UNITS OF MEASURE

The metric system is based on multiples of 10. This makes conversions from a small unit to a larger unit or a larger unit to a smaller unit easy. It can be done by sliding the decimal point to the left or right. Common prefixes are shown in the table.

| Prefix | What it means |
|--------|---------------|
| kilo | one thousand |
| centi | one-hundredth |
| milli | one-thousandth |
| micro | one-millionth |
| nano | one-billionth |

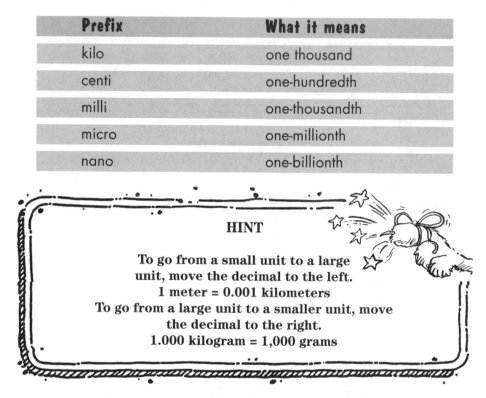

### HINT

To go from a small unit to a large unit, move the decimal to the left.
1 meter = 0.001 kilometers
To go from a large unit to a smaller unit, move the decimal to the right.
1.000 kilogram = 1,000 grams

## COMMON UNITS OF METRIC MEASURE

| Exponent | Decimal | Mass | Length | Volume |
|----------|---------|------|--------|--------|
| $10^3$ | 1000 | kilogram | kilometer | kiloliter |
| $10^0$ | 1 | gram | meter | liter |
| $10^{-2}$ | 0.01 | centigram | centimeter | centiliter |
| $10^{-3}$ | 0.001 | milligram | millimeter | milliliter |
| $10^{-6}$ | 0.000001 | microgram | micrometer | microliter |
| $10^{-9}$ | 0.000000001 | nanogram | nanometer | nanoliter |

# BRAIN TICKLERS
## Set #4

1. A drink bottle contains 250 ml of water. How many liters of water does the bottle contain?

   a. 2.50 liters
   b. 25.0 liters
   c. 0.025 liters
   d. 0.250 liters

2. If the volume of a quarter is 808.93 mm$^3$, what is the volume in cubic centimeters (cm$^3$)?

   a. 0.080893 cm$^3$
   b. 0.80893 cm$^3$
   c. 8.0893 cm$^3$
   d. 80.893 cm$^3$

3. If a metric weight is labeled 1.5 kg and a post-1982 copper penny has a mass of 2.5 g, approximately how many pennies would it take to balance the metric weight?

(Answers are on page 20.)

# MODERN ATOMIC THEORY

In the early 1800s, **John Dalton** reviewed the major discoveries of other scientists and proposed a theory that matter was composed of indivisible particles called atoms. Since that time, research has added to our understanding of the structure of the atom. Although the atom has been broken into smaller subunits, it is still believed to be the fundamental complete building block of matter.

# Characteristics of Atoms

1. Atoms can be broken down into electrons, protons, and neutrons. Protons and neutrons can be further broken down into quarks.
2. In any element, all the atoms have the same number of protons. Atoms of an element with a different number of neutrons are called isotopes.
3. Atoms of different elements are different.
4. Atoms of two or more elements can combine to form compounds.
5. The average atomic weight (mass) of an element reflects the average of all its isotopic forms, but is unique to the element.
6. Atoms of elements in a compound combine in a constant ratio.

# Parts of the Atom

In 1897, English physicist **Joseph John (J. J.) Thomson** (1856–1940) discovered the **electron**. The electron has a negative charge and an extremely tiny mass compared with the mass of an atom. The motion of the electron generates both an electric and a magnetic field. Thomson proposed the plum pudding model of the atom in which negatively charged particles were randomly distributed within a pudding of positively charged particles. Thomson's graduate student Ernest Rutherford disproved the theory.

In 1909, Ernest Rutherford conducted a series of experiments in which he bombarded a thin sheet of gold foil with alpha particles. Some particles passed through the foil, whereas a few were scattered back. From this Rutherford concluded that the **volume of the atom was mostly empty space** and that negative electrons circled around a dense positive core.

In 1919, Rutherford discovered the **proton**, a positively charged particle with a mass approximately 1836 times greater than the mass of an electron. The proton was at the center of the cloud of electrons.

In 1932, English physicist **James Chadwick** discovered the **neutron**, a neutrally charged particle that has a mass slightly greater than that of the proton. Chadwick suggested that there was a strong force that held neutrons and protons together in the nucleus or center of the atom.

| Particle | Charge | Symbol | Location | Mass atomic mass units | Mass grams |
|---|---|---|---|---|---|
| Proton | Positive +1 | ⊕ | Nucleus | 1 | $1.67 \times 10^{-24}$ |
| Neutron | Neutral 0 | ⓪ | Nucleus | 1 | $1.67 \times 10^{-24}$ |
| Electron | Negative −1 | ⊖ | Energy shell or orbital | $\dfrac{1}{1836}$ | $9.11 \times 10^{-28}$ |

Figure 1–6. Proton, Neutron, Electrons

In 1915, Danish physicist **Niels Bohr** (1885–1962) developed an atomic model with electrons organized in energy shells around a nucleus of protons and neutrons. His planetary atomic model is not entirely correct, but it simplifies atomic structure for the beginning chemistry student.

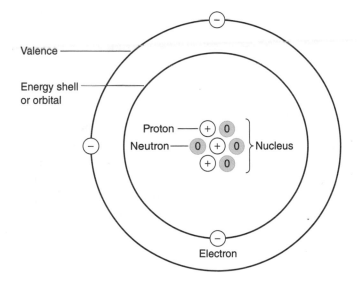

Figure 1–7. Atomic structure

Protons and neutrons are in the core of the atom. They form the **nucleus**. The electrons are organized into energy shells or orbitals. The electrons form a cloud around the nucleus. The last layer of electrons is called the **valence**.

All particles of the atom vibrate. The frequency of vibrations can be used to identify specific atoms and to study the molecular spectra, heat capacity, and heat conduction of materials made from a combination of atoms.

In 1935, Japanese physicist **Hideki Yukawa** (1907–1981) proposed a theory of smaller particles that held the neutrons and protons together in the nucleus. In 1973, **Murray Gell-Mann** (1929– ) led the development of the **quark theory**. There are six types of quarks. Up and down quarks are the fundamental building blocks of protons and neutrons. The sum of the quark charges explains the charges of the proton and neutron.

| Particle | Charge | Symbol | Quarks | Charge Calculation |
|---|---|---|---|---|
| Proton | Positive | (+) | 2 Up 1 Down | $\frac{2}{3} + \frac{2}{3} - \frac{1}{3} = +1$ |
| Neutron | Neutral | (0) | 1 Up 2 Down | $\frac{2}{3} - \frac{1}{3} - \frac{1}{3} = 0$ |

Gluons hold the quarks together.
The strong force holds the nucleus together.

Figure 1–8. Quarks

Like electrical charges repel each other, and unlike electrical charges attract each other. The difference in electrical charges keeps the negative electrons buzzing around the positive nucleus. But what keeps the positive and neutral particles of the nucleus together? This is a different type of attraction called the **strong force**.

# RELATIVE PARTICLE SIZES

The size of the atom and its parts are too small to be seen by the unaided human eye. Scientists who study the structure of the atom use instruments such as the tunneling electron microscope, particle accelerators, and spectroscopes.

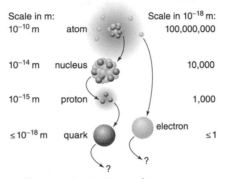

Figure 1–9. Particle sizes

If the nucleus of an atom were the size of a period or 0.5 millimeter, then the first electron would be orbiting at a distance of about 50 meters. If the period were dropped on the fifty-yard line of a football field, then the first electron would be a little more than 4.5 yards into the end zone. If a hydrogen atom were as large as an average apple, then the apple would be as large as the Earth. To keep the math simple, a unit called the angstrom is used.

$$1 \text{ angstrom} = 1.0 \times 10^{-10} \text{ meters} = 1 \text{ nanometer}$$

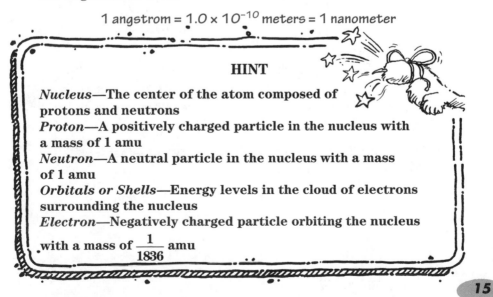

## HINT

*Nucleus*—The center of the atom composed of protons and neutrons

*Proton*—A positively charged particle in the nucleus with a mass of 1 amu

*Neutron*—A neutral particle in the nucleus with a mass of 1 amu

*Orbitals or Shells*—Energy levels in the cloud of electrons surrounding the nucleus

*Electron*—Negatively charged particle orbiting the nucleus with a mass of $\frac{1}{1836}$ amu

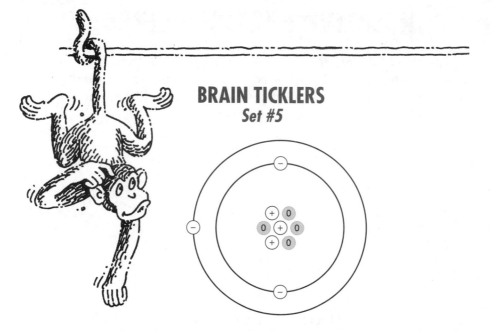

## BRAIN TICKLERS
### Set #5

1. The subatomic particle shown in the diagram above with a positive charge is
   a. a proton
   b. a neutron
   c. an electron
   d. a quark

2. What two subatomic particles contribute most of the mass of the atom?

3. From the diagram, which of the following is a correct inference?
   a. For every positive charge there is a neutral charge.
   b. For every neutral charge there is a negative charge.
   c. For every positive charge there is a negative charge.
   d. None of these statements is correct.

4. What holds the nucleus of the atom together?
   a. electromagnetic force
   b. strong force
   c. gluons
   d. weak force

5. Most of the volume of the atom is composed of
   a. the nucleus
   b. the electrons
   c. empty space
   d. energy shells

6. The particles discovered by J. J. Thomson and placed in orbitals by John Dalton are
   a. neutrons
   b. electrons
   c. quarks
   d. protons

7. If an up quark has a charge of $+\frac{2}{3}$ and a down quark has a charge of $-\frac{1}{3}$, what combination of three quarks would create a neutron? Show the equation.

8. A dime has a mass of 2 grams. If an electron had the same mass as a dime, how much mass would a proton have?
   a. 3,785 grams or about 1 gallon of water
   b. 4.646 kilograms or about 10 pounds of flour
   c. 2,278 grams or about 5 pounds of sugar
   d. 4.0 kilograms or about 1 gallon of milk

(Answers are on pages 20–21.)

## Wrapping Up

- Matter is a substance that is defined by mass and volume.
- Density is mass per volume.
- The metric system is based on powers of 10.
- The modern atomic theory recognizes that the atom can be broken into smaller particles.
- The main components of the atom are the proton, neutron, and electron.
- Protons have a positive charge, a mass of 1 atomic mass unit, and are in the nucleus. Ernst Rutherford discovered the proton.

- Neutrons have no charge, a mass of 1 atomic mass unit, and are in the nucleus. James Chadwick discovered the neutron.
- Protons and neutrons are made of tinier particles called quarks that are held together by gluons. Murray Gell-Mann led the discovery of quarks.
- Electrons have a negative charge and a mass of 1/1836 atomic mass units, and are in energy shells or orbitals. J. J. Thomson discovered the electron.
- The last orbital or energy level is called the valence.
- Most of the volume of the atom is empty space.
- Most of the mass of the atom is in the nucleus.
- In a neutral atom, the protons and electrons occur in a 1:1 ratio.

# BRAIN TICKLERS—THE ANSWERS

## Set #1, page 4

1. The scale is an electronic balance.

2. The mass of 10 marbles is 100 grams. The mass of 1 marble is total mass divided by the number of marbles, or $\frac{100 \text{ grams}}{10 \text{ marbles}}$. The mass of 1 marble is 10 grams.

3. The mass of the brass plate is constant and does not depend on gravity. Therefore the mass of the brass plate would be the same on the moon as on Earth. Gravity would affect the weight of the brass plate, but not the mass.

## Set #2, page 7

1. The change in water volume is the volume of the chunk of granite: 34 ml – 30 ml = 4 ml. The volume of the granite is 4 ml.

2. Volume = length × width × height; therefore, 20 cm × 5 cm × 10 cm = 1000 cm$^3$.

3. The longest side of the box is 20 centimeters, one-fifth of a meter or approximately 7 inches. You could use a metric ruler to measure the dimensions of the box. You could use a meter stick, but it would be like using a yardstick instead of a 12-inch ruler—a little awkward.

4. The volume of the ball can be calculated using the formula for a sphere. The calculated volume is 33.5 cm$^3$. The ball could also be placed in a graduated cylinder of water. The ball will displace a volume of water equal to the volume of the ball.

## Set #3, page 9

1. b.  27 g/ml$\left(\dfrac{5.4 \text{ g}}{2.0 \text{ ml}}\right)$

2. Density is $\dfrac{\text{mass}}{\text{volume}}$; therefore volume is $\dfrac{\text{mass}}{\text{density}}$. The volume of a penny is 0.36 ml.

3. Density of a pure substance is an identifying characteristic. The density of a penny is 6.94 g/cm$^3$. If the penny were pure copper, the density would be 8.96 g/cm$^3$. Therefore, the penny is not pure copper.

## Set #4, page 11

1. d.  0.250 liters

   There are 1,000 ml in a liter.  Slide the decimal three places to convert 250 ml to liters. The answer is 0.250 liter ($\frac{1}{4}$ of a liter).

2. b.  0.80893 cm$^3$

   Centimeters are one power of 10 larger than millimeters. You need to slide the decimal to the left. You need to slide the decimal one space for each time the unit is multiplied. Because the unit is cubed, the decimal moves three places.

3. The number of pennies would need to equal 1.5 kg or 1,500 grams. Divide 1,500 grams by 2.5 grams. The number of pennies needed to balance the metric weight is 600.

## Set #5, page 16

1. a.  a proton

2. Neutrons and protons in the nucleus account for the mass of an atom.

3. c.  For every positive charge there is a negative charge.

4. b. strong force

5. c. empty space

6. b. electrons

7. The charge of a neutron is zero. The combination would be 1 up quark and 2 down quarks.

$$\frac{2}{3} + \left(-\frac{1}{3}\right) + \left(-\frac{1}{3}\right) = 0$$

8. d. 4.0 kilograms or about 1 gallon of milk

$2,000 \times 2 \text{ g} = 4,000 \text{ g} = 4.0 \text{ kg}$

# CHAPTER TWO

# The Periodic Table
# of the Elements

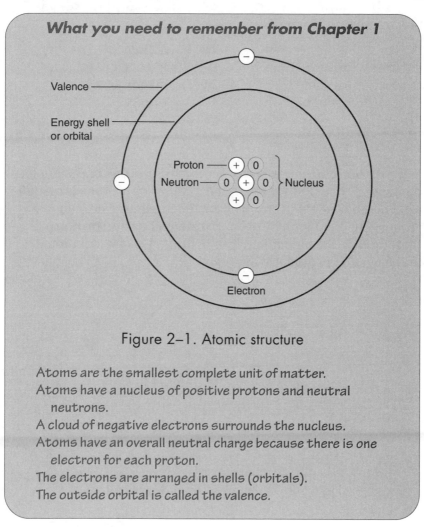

## What you need to remember from Chapter 1

Figure 2–1. Atomic structure

Atoms are the smallest complete unit of matter.
Atoms have a nucleus of positive protons and neutral neutrons.
A cloud of negative electrons surrounds the nucleus.
Atoms have an overall neutral charge because there is one electron for each proton.
The electrons are arranged in shells (orbitals).
The outside orbital is called the valence.

# ELEMENTS

**Elements** are the building blocks of more complex matter. Elements cannot be broken into simpler substances by normal chemical processes. Each element is a unique atom with a specific number of protons in the nucleus, an atomic weight, identifying physical properties, and an electron arrangement that determines its chemical properties. Each element is assigned a one- or two-letter symbol.

Some common elements are

| H | hydrogen | He | helium |
|---|----------|----|--------|
| O | oxygen | N | nitrogen |
| Na | sodium | Mg | magnesium |
| Cu | copper | C | Carbon |

# Atomic Number

The **atomic number** is the number of protons in the nucleus of an atom. In a neutral atom, the number of protons is also the total number of electrons orbiting the nucleus. The simplest atom is the element hydrogen. Hydrogen has one proton; therefore, the atomic number of hydrogen is one. All atoms that have one proton are hydrogen.

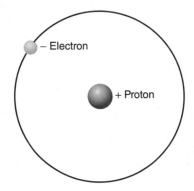

Figure 2–2. Diagram of hydrogen atom

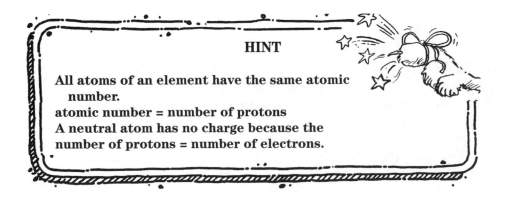

**HINT**

**All atoms of an element have the same atomic number.**
**atomic number = number of protons**
**A neutral atom has no charge because the number of protons = number of electrons.**

# Isotopes of Elements

Although all atoms of an element have the same number of protons, the number of neutrons may vary. Atoms that have the same number of protons but a different number of neutrons are called **isotopes**. Isotopes share the same atomic number, but have a different atomic mass because the mass of a neutron is nearly the same as the mass of a proton (1 atomic mass unit or 1 amu).

The most common form of hydrogen (protium) has 1 proton and no neutrons. The atomic mass is 1 amu. However, an isotope of hydrogen called deuterium has 1 proton and 1 neutron. The atomic mass of deuterium is 2 amu. Another isotope of hydrogen is tritium. Tritium has 1 proton and 2 neutrons and an atomic mass of 3 amu.

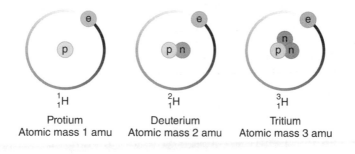

$^{1}_{1}\text{H}$  
Protium  
Atomic mass 1 amu

$^{2}_{1}\text{H}$  
Deuterium  
Atomic mass 2 amu

$^{3}_{1}\text{H}$  
Tritium  
Atomic mass 3 amu

Figure 2–3. Isotopes of hydrogen

Think of building a nucleus using marbles that weigh 1 gram each. Protons could be blue and neutrons red. Protium would have 1 blue marble and a mass of 1 gram. Deuterium would have 1 blue marble and 1 red marble and a mass of 2 grams. Tritium would have 1 blue marble and 2 red marbles and a mass of 3 grams. All of your models are hydrogen because they all have 1 blue marble (1 proton). The difference is the number of red marbles (neutrons).

Notice that the symbol for hydrogen (**H**) has a superscript and a subscript. The superscript is the atomic mass and the subscript is the atomic number of the isotope. The symbol $^{2}_{1}\text{H}$ shows that deuterium has an atomic number of 1 and an atomic mass of 2 amu.

## Atomic Weight (Mass)

The **atomic weight (mass)** of an element is the **average mass** of a sample containing many atoms of the element. The average is compared to the mass of carbon-12. Because it is an average that includes isotopes, the atomic weight of an element (mass) shown on the periodic table is usually not a whole number.

## Atomic Mass Number

**Atomic mass number** is the sum of the number of protons and neutrons in the nucleus of an atom.

Atomic mass number = number of protons + number of neutrons

The mass of the electrons in an atom is ignored in the calculation of atomic mass number, because the mass of the electron is $\frac{1}{1836}$ of a proton. Worrying about the mass of the electrons in the atomic mass of an element is the same as worrying about the mass of your socks when weighing yourself on a scale.

The average mass of a sample of hydrogen atoms is 1.00792 amu. It can be inferred that most hydrogen atoms have one proton and no neutrons, because the atomic mass is close to 1 amu. However, because the average mass is not 1, some atoms of hydrogen have at least one neutron in the nucleus.

To calculate the number of neutrons in the average atom of an element, round the atomic weight to the nearest whole number. Then subtract the atomic number (number of protons).

---

**Sample Calculation**
How many neutrons are in an average atom of chlorine?
Atomic number of chlorine = 17
Atomic Weight = 35.45
Round 35.45 to 35.
Subtract 17 from 35.
35 – 17 = 18
The number of neutrons is 18.

---

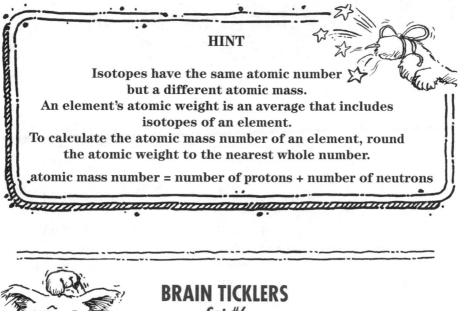

**HINT**

Isotopes have the same atomic number
but a different atomic mass.
An element's atomic weight is an average that includes
isotopes of an element.
To calculate the atomic mass number of an element, round
the atomic weight to the nearest whole number.

atomic mass number = number of protons + number of neutrons

# BRAIN TICKLERS
*Set #6*

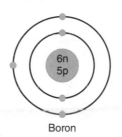

Boron

Figure 2–4. Boron

1. What is the atomic number of boron?

2. The atomic weight of boron is 10.811 amu. Why is the weight not 11 amu?

3. A carbon atom has 6 electrons. What is the atomic number of carbon?

4. The atomic weight of carbon is 12.011. How many neutrons are in the nucleus of a common carbon atom?

5. The symbol for oxygen is O. What does $^{16}_{8}O$ mean? How many protons and neutrons does this isotope of oxygen have?

(Answers are on page 46.)

29

# THE PERIODIC TABLE OF THE ELEMENTS

In the late 1860s, Russian chemist Dmitri Mendeleev organized the 63 known elements into a table. He began by arranging the elements into rows according to increasing atomic mass. Then Mendeleev arranged the elements by groups of elements that had similar properties. There were gaps in the table that Mendeleev predicted would be filled by as yet undiscovered elements. In time the gaps would be filled and more elements would be added.

The modern periodic table of the elements is based on the discovery of atomic structure and electron configuration. Elements are arranged by increasing atomic number. Scientists use the periodic table of the elements to predict chemical reactions of elements with one another and to predict chemical properties of elements. (See the Periodic Table of the Elements on page 43.)

## Metals, Metalloids, and Nonmetals

There are three major categories of elements: metals, metalloids, and nonmetals. A stair-step-like boundary line on the periodic table separates the metals from the nonmetals. **Metals** are the most common type of element and are on the left side of the boundary line. **Metalloids** form the boundary between the metals and the nonmetals. **Nonmetals** are found to the right of the boundary line. The identifying properties of each category are as follows:

|  |  |
|---|---|
| **Metals** | shiny, hard, malleable (can be pounded into sheets or molded into shapes), ductile (can be pulled into a wire), high melting point, good conductor of electricity and heat; solids, except for mercury Examples: gold, silver, aluminum, tin, zinc, iron, copper |
| **Metalloids** | often called semimetals that have characteristics of metals and nonmetals; may be semiconductors, acting as both an |

insulator and a conductor of electricity, depending on conditions

Examples: silicon, germanium, boron, arsenic, antimony, tellurium, polonium

**Nonmetals**    low density, low boiling point, poor conductors of electricity, good insulators, solids are brittle and low luster (not shiny); may be solid, liquid, or gas, most are gases at normal temperature and pressure

Examples: carbon, oxygen, nitrogen, hydrogen

The diagram shows the relative locations of element types on the periodic table of the elements. Note that hydrogen is a nonmetal.

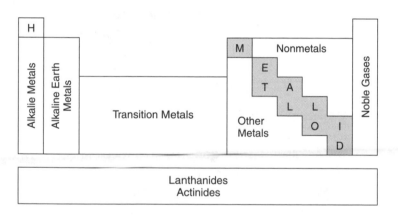

Figure 2–5. Map of periodic table

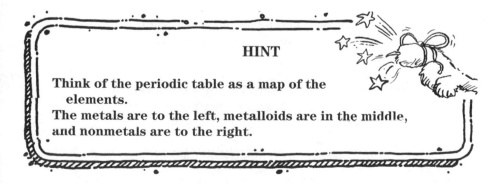

### HINT

Think of the periodic table as a map of the elements.
The metals are to the left, metalloids are in the middle, and nonmetals are to the right.

# BRAIN TICKLERS
## Set #7

1. Mendeleev organized the first periodic table of the elements according to
   a. increasing atomic number
   b. increasing valence electrons
   c. increasing atomic mass
   d. increasing electron shells

2. From an element's location on the periodic table, chemists can predict
   a. its chemical properties
   b. its chemical name
   c. its chemical symbol
   d. its common name

3. Which of the following is a true statement about metal elements?
   a. Metals are often good semiconductors.
   b. Metals are malleable.
   c. Metals are low-density materials.
   d. Metals have low melting points.

4. Relative to the boundary line, where are the metalloids located?

5. An element is a gas at normal temperature and pressure. Is it a metal, nonmetal, or metalloid? Explain your answer.

6. What is a main difference between the modern periodic table and Mendeleev's periodic table of the elements?

(Answers are on page 46.)

# USING THE PERIODIC TABLE OF THE ELEMENTS

## Periods of the Periodic Table

The horizontal rows of the periodic table are called periods. There are seven periods. The trends across a period are as follows:

- The atomic number, atomic mass number, and atomic weight of the elements increase from left to right across a period.
- All elements of a period have the same number of electron shells. The period number is the number of electron shells.
- The valence or last orbital fills across a period.
- Except for period 1, each period begins with a metal and ends with a nonmetal.
- Except for period 1, each period begins with an element that has one valence electron and ends with an element that has 8 valence electrons.
- In periods 2 through 7, metalloids separate the metals from the nonmetals.

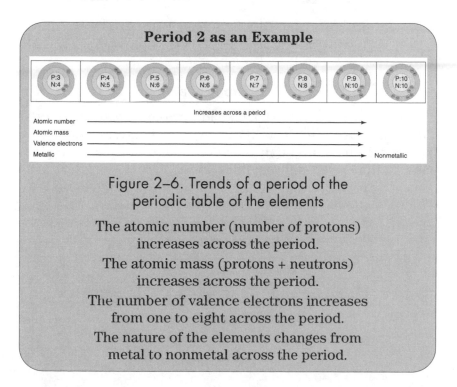

### Period 2 as an Example

Figure 2–6. Trends of a period of the periodic table of the elements

The atomic number (number of protons) increases across the period.

The atomic mass (protons + neutrons) increases across the period.

The number of valence electrons increases from one to eight across the period.

The nature of the elements changes from metal to nonmetal across the period.

Period 1 is the exception because it contains only two elements, hydrogen and helium. These two elements have one electron orbital that can hold up to two electrons. Both hydrogen and helium are nonmetals.

## Electron Arrangement

Most diagrams of an atom are drawn as flat rings surrounding a core. For periods 1–3, these rings or shells fill in an easy-to-understand pattern across a period. Period 1 fills up with two electrons in the first ring. Period 2 fills up with 2 electrons in the first ring and 8 electrons in the second ring. Period 3 fills up with 2 electrons in the first ring, 8 electrons in the second ring, and 8 electrons in the third ring.

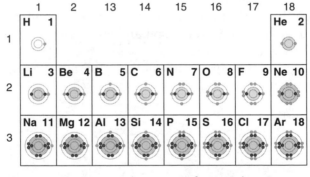

Figure 2–7. Electrons of period 1–3

Electron arrangements become more complicated in periods 4 through 7, because electrons are actually in layers, not nice flat rings. Think of the atom as a never-ending jawbreaker candy with multiple layers surrounding a core.

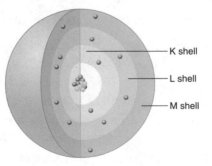

Figure 2–8. Shells

The layers are called shells. Each shell has subshells. The shells are named K, L, M, N, O, P, and Q. The first subshell of each shell is s. Additional subshells are p, d, and f.

Through the transition metals, electrons may fill spaces in the subshells of lower shells before filling the outer shell. This is because electrons tend to spin in pairs.

The rules for filling subshells are complicated and will be covered in advanced chemistry courses. For now, focus on understanding how patterns and trends can be shown on the periodic table.

The diagram shows the electron arrangements of the group 13 elements. Notice that the electrons fill lower energy levels and that all members of the group have 3 valence electrons.

**Elements of group 13**

| Period shell | Electron arrangement | Element name / Electron configuration |
|---|---|---|
| 2 L | P:5 N:6 | Boron 2, 3 $s^2, s^2p^1$ |
| 3 M | P:13 N:14 | Aluminum 2, 8, 3 $s^2, s^2p^6, s^2p^1$ |
| 4 N | P:31 N:39 | Gallium 2, 8.18, 3 $s^2, s^2p^6, s^2p^6d^{10}, s^2p^1$ |
| 5 O | P:49 N:66 | Indium 2, 8, 18, 18, 3 $s^2, s^2p^6, s^2p^6d^{10}, s^2p^6d^{10}, s^2p^1$ |
| 6 P | P:81 N:123 | Thalium 2, 8, 18, 32, 18, 3 $s^2, s^2p^6, s^2p^6d^{10}, s^2p^6d^{10}f^{14}, s^2p^6d^{10}, s^2p^1$ |

Figure 2–9. Group 13 elements

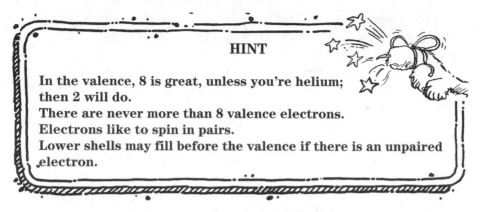

**HINT**

In the valence, 8 is great, unless you're helium;
then 2 will do.
There are never more than 8 valence electrons.
Electrons like to spin in pairs.
Lower shells may fill before the valence if there is an unpaired
electron.

# Blocks of the Periodic Table

How electrons spin, pair up, and fill energy levels explains the
properties of metals such as magnetism, electrical conductivity,
and luster, the semiconducting property of metalloids, and the
chemical properties of nonmetals. **Blocks** of the periodic table
are regions of elements that demonstrate similar electron filling
trends across a period. The diagram shows the general trends
across the periods.

**Electron shell filling trends**

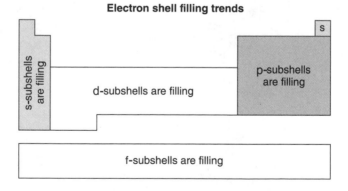

Figure 2–10. Electron shell filling trends

An element of the first period elements has one energy shell,
K, with one s-subshell. Elements of the second period have a
full K shell and fill the L shell's s-subshell and then the p-
subshell. The pattern continues with each period adding
another shell and filling subshells across the period.

The electron configuration is often shown on the periodic
table of the elements. For example, the electron configuration

of copper is 2-8-18-1. The diagram shows the arrangement of the electrons as 2 electrons in the K shell, 8 electrons in the L shell (2 in the s, 6 in the p), 18 electrons in the M shell (2 in the s, 6 in the p, 10 in the d), and 1 electron in the N orbital.

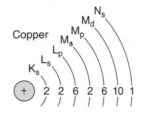

Figure 2–11. Electron arrangement of copper

The good news is that most of the time in chemistry you are concerned about the number of electrons in the valence and ignore the other electrons.

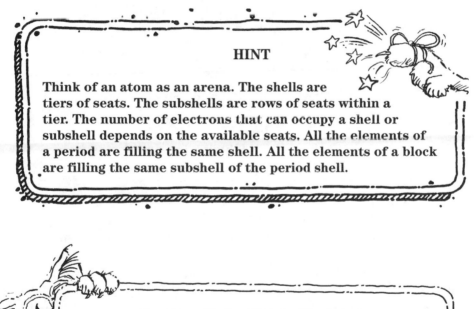

**HINT**

Think of an atom as an arena. The shells are tiers of seats. The subshells are rows of seats within a tier. The number of electrons that can occupy a shell or subshell depends on the available seats. All the elements of a period are filling the same shell. All the elements of a block are filling the same subshell of the period shell.

**Caution—Major Mistake Territory!**

Under certain conditions, when forming bonds with other atoms, the valence of some elements in the d-block may include electrons from a lower subshell.

# Groups and Families of the Periodic Table

| Block | Groups and Families |
|-------|---------------------|
| s | Hydrogen and Helium |
| | Group 1 Alkali Metals |
| | Group 2 Alkaline Earth Metals |
| d | Groups 3–12 Transition Metals |
| p | Group 13 Boron family metalloids and poor metals |
| | Group 14 Carbon family of nonmetals, metalloids, and poor metals |
| | Group 15 Nitrogen family of nonmetals, metalloids, and poor metals |
| | Group 16 Oxygen family of nonmetals, metalloids, and poor metals |
| | Group 17 Halogens |
| | Group 18 Noble Gases |
| f | Rare Earth Metals—Lanthanides and Actinides |

Columns of the periodic table form vertical **groups** of elements that have the same number of valence electrons. Some groups have such similar properties that they are given **family** names.

Groups 1 and 2 fill the s-subshell; therefore, they are known as the s-block of the periodic table. The elements of groups 1 and 2 are soft metals. Pure sodium, for example, can be cut with a plastic knife. The metals of groups 1 and 2 readily form salts with nonmetals of group 17, the halogen family.

## Group 1

Group 1 is known as the family of **alkali metals** and includes the elements lithium, sodium, potassium, rubidium, cesium, and francium. All of the alkali metals are highly reactive and have 1 valence electron. Alkali metals are rarely found in pure form in nature. Lithium and sodium react violently with water and are stored in oil.

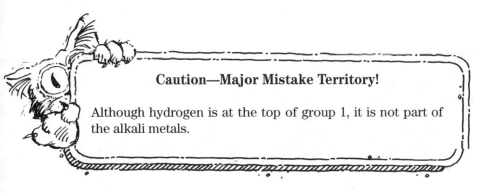

**Caution—Major Mistake Territory!**

Although hydrogen is at the top of group 1, it is not part of the alkali metals.

## Group 2

Group 2 is known as the family of **alkaline earth metals** and includes the elements beryllium, magnesium, calcium, strontium, barium, and radium. The alkaline earth metals have 2 valence electrons. Magnesium burns with a bright white light and strontium has a deep red flame. Both are used in fireworks.

## Groups 3–12

Groups 3 through 12 are the **transition metals**. These are the elements that are generally shiny, hard, dense, and good conductors of electricity and heat. There are no families in these groups. They form the d-block of the periodic table because they are filling the d-subshell across each period. Because the electrons are filling lower energy levels across the period, there is no memory trick for correlating the group number and the number of valence electrons. However, all elements of a group have the same number of valence electrons.

## Groups 13–16

Groups 13 through 16 are **mixed groups** that may contain metalloids, metals, and nonmetals. They form the p-block of the periodic table because they fill the p-subshell. All elements of a group share the same number of valence electrons. For groups 13 (boron family), 14 (carbon family), 15 (nitrogen family, also known as the pnictogens), and 16 (oxygen family, also known as chalogens), the number of valence electrons is the last digit of the group number. For example, all elements of group 13 have 3 valence electrons.

## Group 17

Group 17 is known as the family of **halogens** and includes fluorine, chlorine, bromine, iodine, and astatine. All of the halogens are nonmetals. All the elements of group 17 have 7 valence electrons. Halogens are often found combined with metals of groups 1 and 2—for example, sodium chloride, better known as table salt.

## Group 18

Group 18 is known as the family of **noble gases**. The gases of group 18 have a full valence and are **inert** or not reactive. They do not need to share, gain, or lose electrons. The noble gases are helium, neon, argon, krypton, xenon, and radon. Helium has two valence electrons. The rest of the group 18 elements have 8 valence electrons.

## Rare Earth Metals

Notice that there are two rows of elements at the bottom of the periodic table. These are the **rare earth metals** known as the **lanthanides** (elements 57–70) and **actinides** (elements 89–102). Uranium is an example of a rare earth metal. The lanthanides and actinides form the f-block of the periodic table because the f-subshell is filling. They are separated from the rest of the table for convenience.

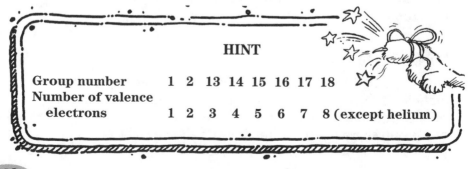

**HINT**

| Group number | 1 | 2 | 13 | 14 | 15 | 16 | 17 | 18 |
|---|---|---|---|---|---|---|---|---|
| Number of valence electrons | 1 | 2 | 3 | 4 | 5 | 6 | 7 | 8 (except helium) |

# BRAIN TICKLERS
### Set #8

1. Which of the following is true of the elements of groups of the periodic table?
   a. The elements share the same number of electron shells.
   b. The elements are arranged by decreasing atomic mass.
   c. The elements share similar properties.
   d. The elements are the same phase of matter.

2. Which of the following is NOT true of periods of the periodic table?
   a. The atomic number increases from left to right across a period.
   b. The atomic mass increases from left to right across a period.
   c. The number of valence electrons increases from left to right across a period.
   d. The metallic properties of elements increase from left to right across a period.

3. Where are the most reactive metals found in the periodic table?

4. Which elements have two valence electrons?

5. What is the family name of elements of group 17?

6. What family of elements is nonreactive because the valence of each family member is full?

7. A metal reacts violently with water. In which group of the periodic table would it likely be found?

8. In which block of the periodic table are transition metals found?

(Answers are on pages 46–47.)

# Using the Periodic Table

Each block of the periodic table displays information about an element that can be used to predict bonding to form molecules and compounds. The phase of the element at normal temperature and pressure is sometimes also shown on the block.

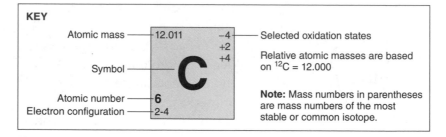

Figure 2–12. Key block

According to the block, carbon has 6 protons. The electrons are filling the L shell with 2 in the s-subshell and four in the p-subshell. Therefore carbon is in period 2 of the periodic table. There are 4 electrons in the valence. Therefore carbon is in Group 14.

Because the atomic mass is 12.011, there are isotopes of carbon. By rounding the atomic mass to the nearest whole number, 12, and subtracting the 6 protons, it can be inferred that there are 6 neutrons in the average carbon atom.

The oxidation state is used to predict the behavior of an element when an atom of the element becomes a charged particle called an ion. Elements form compounds when atoms lose their neutral charge and become ions. (See Chapter 6.)

## Periodic Table of the Elements

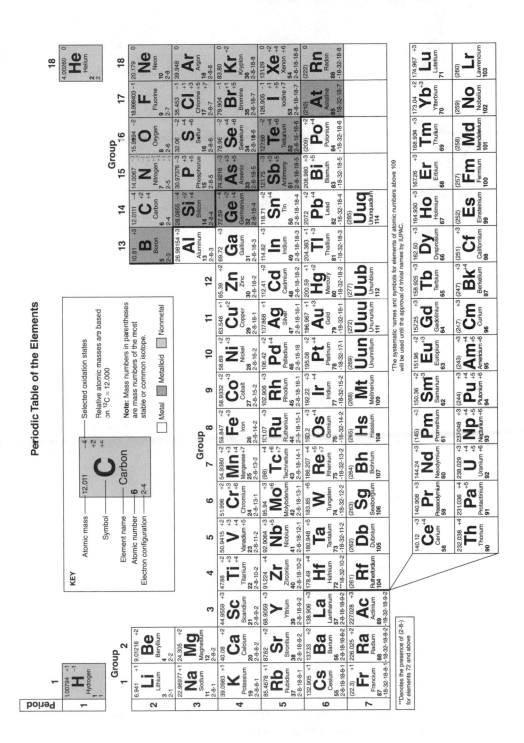

# BRAIN TICKLERS
*Set #9*

*Use the periodic table to answer the following questions.*

1. What is the atomic number, atomic mass, and chemical symbol of copper?

2. What category of elements occupies most of the periodic table?

3. Why are there only two elements in the first period of the periodic table?

4. Of the following metals, magnesium, sodium, and iron, which would most likely be used in construction materials? Explain your answer.

5. How many valence electrons does carbon have?

(Answers are on page 47.)

## Wrapping Up

- An element is composed of a single type of atom with a unique atomic number.
- The atomic number is the number of protons in the nucleus.
- All atoms of an element have the same atomic number.
- The number of electrons is the same as the number of protons.
- The atomic mass is the total mass of the protons and neutrons in the nucleus of an atom.
- The atomic mass of an element is the average mass of a sample of atoms of an element.

- Isotopes of an element have the same number of protons but a different number of neutrons.
- The three major types of elements are metals, metalloids, and nonmetals.
- Metals are good conductors of electricity and heat, have high density and high melting points, and are ductile and malleable.
- Metalloids are semimetals with some properties of metals and some properties of nonmetals. Some are semiconductors.
- Nonmetals are poor conductors of electricity, have lower density and lower melting points than metals, and are brittle in the solid phase.
- Periods are rows of the periodic table of the elements.
- The atomic number increases from left to right across a period.
- The atomic mass increases from left to right across a period.
- The electron valence fills from left to right across a period.
- Groups are columns of the periodic table of the elements.
- Elements of a group have the same number of valence electrons.
- Elements of a group share similar chemical properties.

# BRAIN TICKLERS—THE ANSWERS

## Set #6, page 29

1. The atomic number of boron is 5—the number of protons.

2. Some atoms of boron may have less than 6 neutrons. There are isotopes of boron.

3. The atomic number of carbon is 6. In a neutral atom, there are the same number of electrons as protons.

4. The common carbon atom has a mass of 12 and an atomic number of 6. Therefore, the common carbon atom would have 6 neutrons in the atom.

5. An atom of $^{16}_{8}O$ has an atomic number of 8 and an atomic mass of 16 amu. The atom has 8 protons and 8 neutrons.

## Set #7, page 32

1. c. increasing atomic mass

2. a. its chemical properties

3. b. Metals are malleable.

4. Metalloids are located along the boundary line.

5. A gas has low density. The element is probably a nonmetal.

6. The modern periodic of the table of the elements is arranged by increasing atomic number instead of increasing atomic mass.

## Set #8, page 41

1. c. The elements share similar properties.

2. d. The metallic properties of elements increase from left to right across a period. Why? Metallic properties decrease from left to right.

3. Group 1, the alkali metals

4. Group 2, the alkaline earth metals

5. The halogens

6. Group 18, the noble gases

7. The metal is probably an alkali metal of group 1.

8. Transition metals are the d-block of the periodic table.

## Set #9, page 44

1. The atomic number of copper is 29, the atomic mass is 63.546, and the symbol is Cu.

2. Most of the elements of the periodic table are metals.

3. The K shell can hold only two electrons; therefore the valence is filled after two elements.

4. Iron is a transition metal that would be malleable and ductile. Magnesium and sodium are highly reactive and would not make good building materials.

5. Carbon belongs to group 14 and has 4 valence electrons.

# CHAPTER THREE

# Phases of Matter

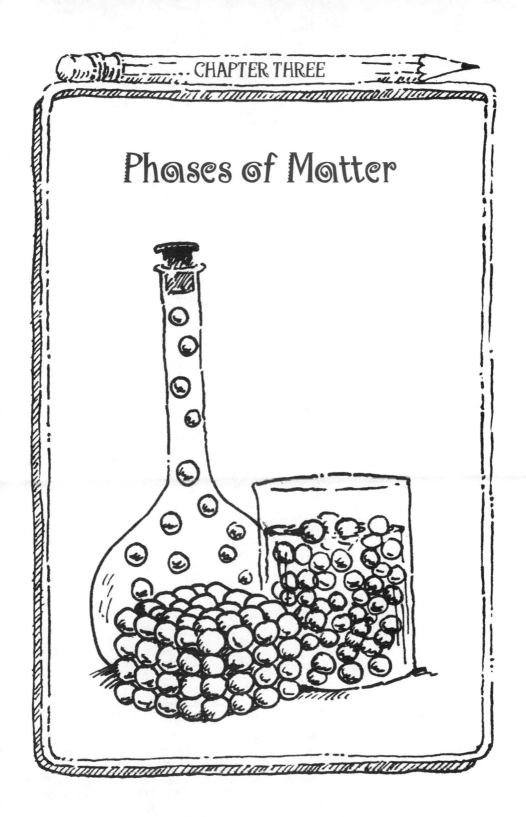

# PHASES OF MATTER

Matter exists in four phases: solid, liquid, gas, or plasma. On Earth, the three most common phases or states of matter are solid, liquid, and gas. The **phase** depends on the kinetic energy of the atoms in a substance. The kinetic energy of a substance depends on temperature and pressure.

**Temperature** is a measure of average kinetic energy. **Pressure** is a force on a given area of a substance. Changes in temperature and pressure can speed up or slow down the movement of atoms in a substance.

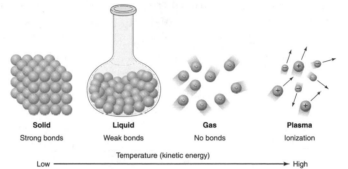

| Solid | Liquid | Gas | Plasma |
|---|---|---|---|
| Strong bonds | Weak bonds | No bonds | Ionization |

Temperature (kinetic energy)

Low ————————————————————→ High

Figure 3–1. Phases of matter and kinetic energy

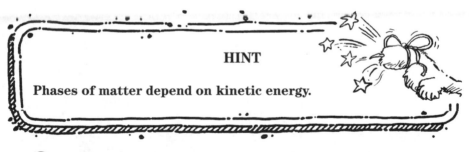

**HINT**

Phases of matter depend on kinetic energy.

# Solids

**Solids** are atoms in an energy state so low that they maintain a shape, occupy a definite volume, and have an identifying density. Solids tend to be tightly packed molecules or atoms that are strongly bonded or attracted to one another.

Although solids appear to be motionless, atoms in the solid are vibrating. Under normal conditions, the volume of a rigid solid does not decrease under pressure. The solid cannot be forced into a smaller space.

Examples of elements that are solids at normal temperature and pressure are metals (except mercury), metalloids, and nonmetals such as carbon, phosphorus, and sulfur. A steel paper clip is an example of a common solid material. Steel is a metal alloy or mixture of iron and other metals.

## Liquids

**Liquids** are atoms in a more excited energy level than solids, but lower energy level than gases. Because the atoms or molecules are weakly bonded or attracted to each other, liquids move as fluids, take the shape of their containers, occupy a definite volume, and have an identifying density or specific gravity at a given temperature.

Like solids, under normal conditions the volume of a liquid does not decrease under pressure. If a force is pushing down on the liquid, the liquid will move out of the way or push back, but the liquid will not be compressed under normal pressures and temperatures.

Examples of elements that are liquids at normal temperature and pressure are the metal mercury and the nonmetal bromine. Glass is an example of a common liquid material; it is a slow-moving fluid. Glass is made from silicon, oxygen, and other elements.

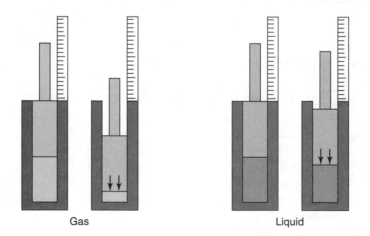

Gas             Liquid

Figure 3–2. Compression of gas and liquid

## Gases

**Gases** are fluids, but the molecules are more spread out than the molecules of a liquid. Gases are atoms or molecules that have high kinetic energy, no definite volume, and no definite shape.

The volume of a gas depends on the size of the container and pressure. Gases can be compressed into a smaller space. Because the volume, pressure, and temperature of a gas are interrelated variables, the density of a gas is given at a specified temperature and pressure. (More on gases in Chapter 4.)

Common elements that are gases at normal temperature and pressure are the nonmetals hydrogen, nitrogen, and oxygen, the halogens (except bromine), and the noble gases. Methane, a compound of carbon and hydrogen, is used in gas stoves and ovens and is an example of common material in the gas phase.

## Plasma

**Plasma** is a gas with such a high kinetic energy that it becomes ionized. Like a gas, plasma has no definite volume, shape, or density. Plasmas can be compressed and unlike gases will react to a magnetic field.

Plasma is the most common phase in the universe. It is found in stars, solar wind, and space between galaxies. On Earth, plasma can be seen as flames, lightning, and the polar auroras. Practical uses for plasma include plasma televisions and fluorescent lamps.

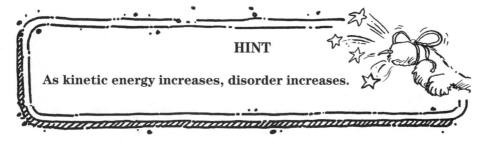

**HINT**

As kinetic energy increases, disorder increases.

# BRAIN TICKLERS
## Set #10

1. In which phase of matter would particles of a substance spread apart and fill a space?
   a. liquid
   b. gas
   c. solid
   d. crystal

2. Which of the following is true about liquids?
   a. Molecules of a liquid condense when the liquid is heated.
   b. Atoms of a liquid are tightly packed together into a definite shape.
   c. Particles of a liquid move freely but maintain weak contact with one another.
   d. The volume of a liquid can be decreased by pressure.

3. An ice cube is an example of what phase of matter?

4. What phase of matter is a stream of ions?

(Answers are on page 69.)

# PHASE CHANGES

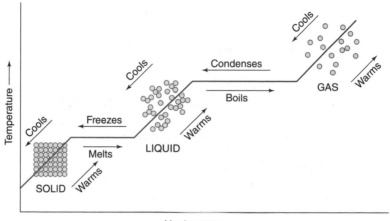

Figure 3–3. Phase changes

## Interpreting the Graph

When a solid warms, it reaches the **melting point** or transition to a liquid phase. As a liquid cools, it reaches the **freezing point** or transition to a solid. The freezing point and melting point are the same temperatures.

When a liquid warms, it reaches the **boiling point** or transition to a gas. As a gas cools, it reaches the **condensation point** or transition to a liquid. The boiling point and condensation point are the same temperatures.

Notice that the temperature is steady during a phase change. During a phase change molecules are a mixture of solid and liquid or liquid and gas. The temperature begins to rise after molecules have made the transition to a single phase.

## Endothermic Phase Changes

During warming, heat energy is added to a substance, the average kinetic energy (temperature) increases, and a phase change may occur. The phase change is **endothermic** if substance absorbs energy from its surroundings.

### Examples of Endothermic Phase Changes

- **Melting**   change from solid to liquid
- **Boiling**   change from liquid to vapor
- **Ionization**   change from gas to plasma

## Exothermic Phase Changes

During cooling, a substance releases heat to its surroundings, the average kinetic energy of the substance decreases, and a phase change may occur. When energy is released to the environment, the phase change is **exothermic**.

### Examples of Exothermic Phase Changes

- **Deionization**   change from plasma to gas
- **Condensation**   change from vapor to liquid
- **Freezing**   change from liquid to solid

As a solid substance warms, the molecules move faster. The **melting point** of a substance is the temperature at which a solid becomes a liquid. When the temperature of a liquid is raised to the **boiling point**, the liquid becomes a gas. As a gas cools, the molecules **condense** into a liquid. When a liquid is cooled to the **freezing point**, a phase change to solid occurs.

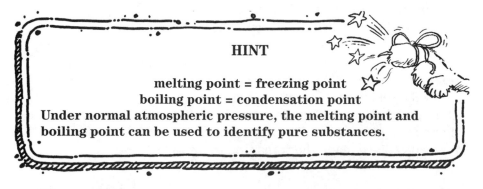

HINT

melting point = freezing point
boiling point = condensation point
Under normal atmospheric pressure, the melting point and boiling point can be used to identify pure substances.

## Evaporation

A liquid may become a gas below the boiling point through **evaporation**. Evaporation occurs at the surface. How fast evaporation takes place depends on wind speed, humidity, surface area, attraction that liquid molecules have for each other, and air temperature.

As liquid molecules leave a moist surface (wet towel, lake, your skin after swimming), the surface is cooled. Evaporation takes energy from the environment; it is endothermic. The opposite process, condensation (gas to liquid), returns energy to the environment; it is exothermic.

## Sublimation and Deposition

**Sublimation** is the term that describes the change from solid to gas without passing through liquid phase. Dry ice, solid carbon dioxide, will change to a vapor through sublimation. On a cold, sunny day, snow will sublimate without melting into a liquid.

The opposite of sublimation is deposition. **Deposition** occurs when a gas enters the solid phase without becoming a liquid. This is how dry ice is made.

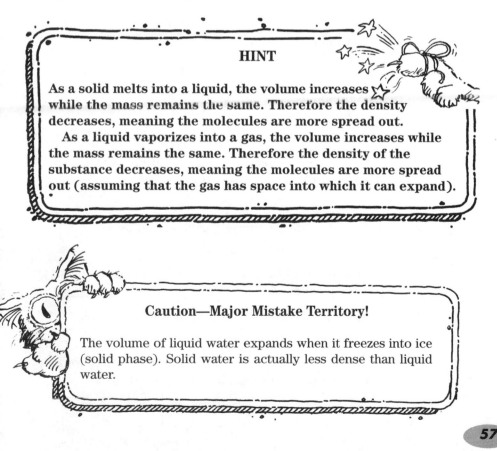

**HINT**

As a solid melts into a liquid, the volume increases while the mass remains the same. Therefore the density decreases, meaning the molecules are more spread out.

As a liquid vaporizes into a gas, the volume increases while the mass remains the same. Therefore the density of the substance decreases, meaning the molecules are more spread out (assuming that the gas has space into which it can expand).

**Caution—Major Mistake Territory!**

The volume of liquid water expands when it freezes into ice (solid phase). Solid water is actually less dense than liquid water.

# BRAIN TICKLERS
## Set #11

1. Describe what would happen to the kinetic energy of a liquid as it is cooled below the freezing point.

2. What tool would be used to measure the change in kinetic energy as a substance undergoes a phase change from liquid to solid?
   a. a ruler
   b. a pan balance
   c. a thermometer
   d. a calculator

3. A patch of ice disappeared during a sunny winter day without forming a wet spot or puddle on the sidewalk. What probably occurred?
   a. melting
   b. sublimation
   c. boiling
   d. freezing

4. A pan of water is placed on a stove burner. The liquid begins to bubble and release steam. The liquid is at its _____ point.

5. Describe the heat exchange that takes place when a liquid becomes a solid.

6. Explain why water droplets form on a bathroom mirror after a hot shower.

7. Explain why fog rises above a pond on a cool fall morning.

8. Describe what would happen if a bottle of water were filled to the top and placed in a freezer.

(Answers are on page 69.)

# READING A PHASE CHANGE GRAPH

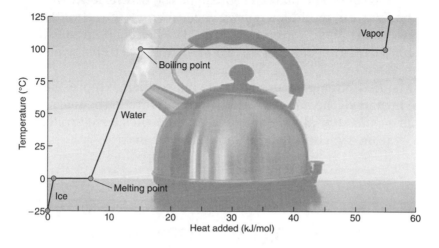

Figure 3–4. Water phase change graph

The temperature of a sample of water was raised from –25°C to 100°C by adding heat. As heat was added, the temperature of the water increased. The data is represented on the graph. Notice that the graph is not a straight line. There are two pauses or plateaus (flat lines with no slope). These plateaus occur at the phase change temperatures of the melting point and the boiling point.

At the start, the water is in the solid phase—ice. Heat energy is added. From –25°C to 0°C, the water molecules gain kinetic energy, but the phase change does not take place immediately.

At the plateau, the molecules are a mix of solid and liquid water. Surface molecules are melting and refreezing until enough energy is added to keep the molecules in the liquid phase.

Additional energy is needed to break the attraction between water molecules in the solid phase. This heat energy is the **latent heat of fusion**. Latent means "hidden." In other words, because the temperature of the water is not changing, it is difficult to tell that heat energy is being added. According to the graph the melting point of water is 0°C.

Between 0°C and 100°C, liquid water absorbs energy. The attraction between the water molecules weakens as the kinetic energy increases. Steam becomes visible.

Steam is a mixture of liquid water and water vapor (water in the gas phase). The plateau represents the **latent heat of vaporization**, or hidden energy used to break the attraction between the liquid water molecules. The boiling point of water is 100°C. All molecules of the water sample are transformed into vapor by the end of the phase change.

The rise in temperature after the phase change represents the heating of the air and water vapor mixture in the teakettle. This occurs after all the liquid water molecules have broken away from each other and entered the gas phase.

# BRAIN TICKLERS
## *Set #12*

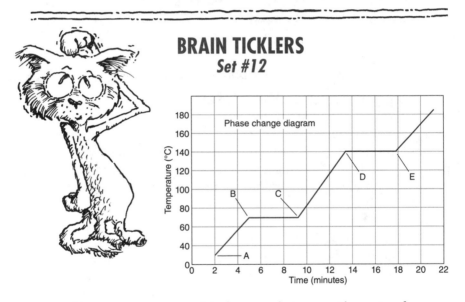

Heat energy was added to a substance. The rate of temperature change is shown on the graph. Answer the questions using information from the graph.

1. What process is occurring between points B and C?
   a. sublimation
   b. freezing
   c. evaporation
   d. melting

2. What phase is the substance between points C and D?
   a. plasma
   b. liquid
   c. gas
   d. solid

3. At what point on the graph did the substance completely vaporize into a gas—A, B, C, D, or E?

4. How long did it take for the substance to change from a solid to a gas?
   a. 18 minutes
   b. 13 minutes
   c. 9 minutes
   d. 21 minutes

5. Heat energy was added to the substance between points B and C. Why did the temperature not change for 4 minutes?

6. Which of the following unknowns is probably the substance shown on the graph?

| Substance (°C) | Melting Point (°C) | Boiling Point (°C) |
|---|---|---|
| 1 | 35 | 70 |
| 2 | 70 | 140 |
| 3 | 80 | 180 |
| 4 | 35 | 140 |

7. Is this substance water? Explain your answer.

(Answers are on pages 69–70.)

# THINGS THAT AFFECT A PHASE CHANGE

In a pure liquid substance, molecules will escape from the surface of a liquid and condense back into a liquid. When the number of molecules escaping from the surface of a liquid equals the number of molecules returning to the liquid phase, a balance or equilibrium is reached. This is called the **equilibrium vapor pressure**.

The vapor pressure of a liquid depends on variables such as temperature, atmospheric pressure, and the attraction that the molecules of a liquid have for each other. Mixing substances into a liquid to form a solution also affects vapor pressure.

## Temperature

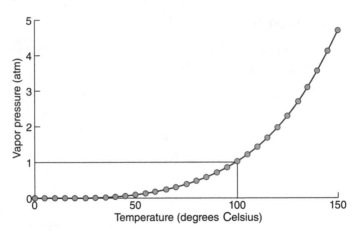

Figure 3–5. Vapor pressure depends on temperature

Vapor pressure depends on temperature. As the temperature of the liquid increases, the vapor pressure increases. The graph shows the vapor pressure of liquid water at increasing temperature. This is not a linear (straight line) relationship. Notice that the graph has a slight positive curve. The relationship is said to be exponential.

Because evaporation takes place at the surface, surface area will affect the rate of vaporization. The evaporation rate increases as surface area increases.

# Atmospheric Pressure and the Boiling Point

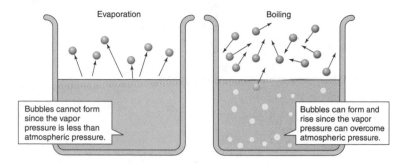

Figure 3–6. Vapor pressure illustration

The **boiling point** of a liquid is the temperature at which its vapor pressure is equal to the pressure of the gas above the liquid. This means that molecules of the liquid are able to overcome their attraction to one another and escape as vapor into the gas above the liquid.

Normally the boiling point of a liquid is measured at 1 atmosphere of pressure. If the pressure above the liquid is lower than 1 atmosphere, then the boiling point will be lower. If the pressure above the liquid is higher than 1 atmosphere, then the boiling point of the liquid will be higher.

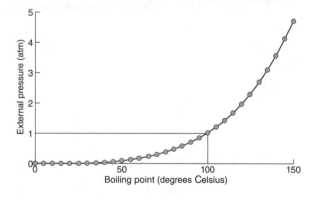

Figure 3–7. Pressure and boiling point curve

According to the graph, the boiling point of water at 1 atmosphere of pressure is 100°C. At 5 times the atmospheric pressure, water boils at 150°C. With extremely low atmospheric pressure, water would boil at temperatures near the normal freezing point!

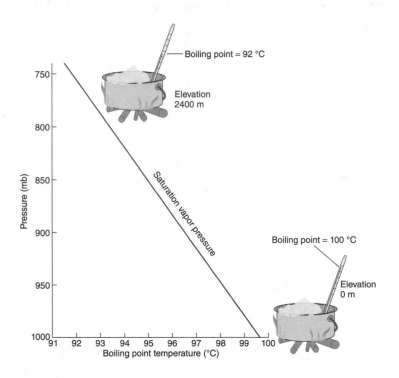

Figure 3–8. Pressure, altitude, and boiling point

In the troposphere, the part of the atmosphere in which you live, air pressure decreases as altitude increases. At high altitudes, liquids boil at a lower temperature than the sea level boiling point. The boiling point of water is 100°C at sea level and 1 atmosphere of pressure (~1,000 millibars). At 2,400 meters altitude, the external pressure is 75% of the pressure at sea level and the boiling point is 92°C.

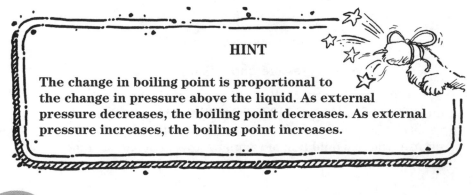

**HINT**

The change in boiling point is proportional to the change in pressure above the liquid. As external pressure decreases, the boiling point decreases. As external pressure increases, the boiling point increases.

# Type of Molecule

The attractions between molecules are called the **inter-molecular forces** or Van der Waals forces. Some molecules are strongly attracted to one another, whereas other molecules are weakly attracted to one another.

If the attraction between molecules is relatively strong, the boiling point will be relatively high, because it takes more energy to pull the molecules away from each other. If the attraction between molecules is relatively weak, then the boiling point will be relatively low, because it takes less energy to pull the molecules away from each other.

# Solutions

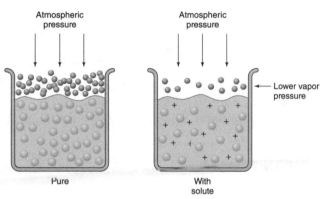

Figure 3–9. Vapor pressure and solutions

**Solutions** are mixtures of substances that are so well mixed, they appear to be one substance. The substance in the greatest quantity is called the **solvent**. The substance in less concentration is the **solute**. The molecules of the solvent and solute interact with one another. (More about solutions in Chapter 5.)

If the solvent is water and solute is salt, the salt molecules will hold the water molecules in the liquid phase. Fewer water molecules will escape from the surface through evaporation. This lowers the vapor pressure above the liquid. The result is that the freezing point is lower and the boiling point is higher. More energy is needed to overcome the forces between the solute and the solvent than is needed to separate the molecules of the pure liquid.

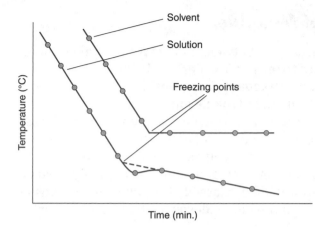

Figure 3–10. Freezing point depression of solution

**HINT**

Adding a solute to a solvent lowers the freezing point and raises the boiling point.
This is called freezing point depression and boiling point elevation.

## BRAIN TICKLERS
### Set #13

1. Death Valley, California, is 85.95 meters below sea level. What will happen to the boiling point of water in Death Valley relative to the boiling point at sea level? Explain your answer.

2. A pan of water was covered and brought to a boil. The water boiled in less time than an equal amount of water in an uncovered pan. Why did this happen?

3. When ice forms on a sidewalk, a thin layer of liquid water forms at the surface if the air is at or slightly below the freezing point of water. Salt thrown on the sidewalk causes the ice to melt. Why does salt cause ice to melt?

4. When a pan of water was placed on a gas stove burner, less heat was needed to bring the water to a boil when the fan over the stove was turned on. Why did the water need less heat when the fan was on?

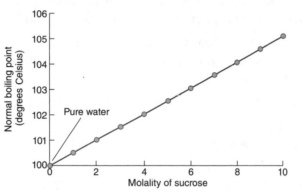

**Effect of Sucrose Concentration on the Boiling Point of Water**

5. Why does the boiling point of water increase as the concentration (molality) of sucrose (sugar) increases?

6. Sketch the slope of the line for the freezing point graph for the water and sugar mixture.

(Answers are on pages 70–71.)

## Wrapping Up

- The phases of matter are solid, liquid, gas, and plasma.
- Solids have a definite shape and fixed volume.
- The molecules of a solid have low kinetic energy.
- Liquids take the shape of their containers and have a fixed volume.

- The molecules of a liquid have a weak attraction and more kinetic energy than a solid.

- Gases have no definite shape or volume.

- The molecules of a gas have higher kinetic energy and no attraction to each other.

- Plasmas have no definite shape or volume.

- The ions of plasma have extremely high kinetic energy and respond to a magnetic field.

- Phases changes depend on heat transfers between a substance and its surroundings.

- Phase changes that are endothermic take energy from the environment and include melting, boiling, and ionization.

- Phase changes that are exothermic give energy to the environment and include deionization, condensation, and freezing.

- Melting transforms a solid to a liquid. Freezing is the reverse of melting.

- Boiling transforms a liquid to a gas. Condensation is the reverse of boiling.

- Ionization transforms a gas to plasma. Deionization is the reverse of ionization.

- Evaporation occurs at the surface of a liquid, transforms water to vapor below the boiling point, and absorbs energy from the environment.

- Sublimation is the transformation of a solid to a gas without a liquid phase. Deposition is the reverse.

- The melting point and boiling point of a substance at a specified temperature and pressure can be used to identify a pure substance.

- The boiling point and freezing points of substances depend on temperature, pressure, type of molecule, and purity of the substance.

# BRAIN TICKLERS—THE ANSWERS

## Set #10, page 54

1. b. gas

2. c. Particles of a liquid move freely but maintain weak
   contact with one another.

3. Solid

4. Plasma

## Set #11, page 58

1. The temperature of the liquid will decrease. The kinetic
   energy will decrease until the liquid becomes a solid.

2. c. a thermometer

3. b. sublimation

4. The liquid is at its <u>boiling</u> point.

5. As the liquid is cooled, it will lose energy to the
   environment.

6. Water vapor in the air condenses on the cooler surface of
   the mirror, becoming liquid water.

7. Water evaporates from the surface of the pond and then
   condenses into fog above the pond.

8. Water expands when it freezes. The ice will push up out of
   the top of the bottle or push out the sides of the bottle. The
   bottle might break if the pressure is too great.

## Set #12, pages 60–61

1. d. melting

2. b. liquid

3. E

4. a.  18 minutes

5. When the substance warmed to the melting point of 70°C, the heat energy was absorbed to break the attraction between the molecules of the substance. This is the latent heat of vaporization.

6. Substance 2 has a melting point of 70°C and a boiling point of 140°C.

7. The substance is not water because the melting point is 70°C, not 0°C, and the boiling point is 140°C, not 100°C.

## Set #13, pages 66–67

1. The boiling point of water in Death Valley will be higher than the boiling point at sea level because the pressure above the water surface will be higher.

2. Covering the pan traps water molecules in the vapor phase, raising the vapor pressure. Raising the vapor pressure above the water in the pan raises the temperature of the water. When the lid is removed, the pressure is reduced and the water comes to a seemingly instant boil.

3. There is a thin layer of water at the surface of the ice. The salt dissolves into the water, lowering the freezing point of the water. With a lower freezing point, the water will stay in the liquid phase.

4. The fan lowered the pressure above the water so that the water boiled at a lower temperature. This required less heat energy to be added to the pan of water.

5. The sugar goes into solution in the water. The sugar keeps the water from leaving the solution. This raises the boiling point.

6.

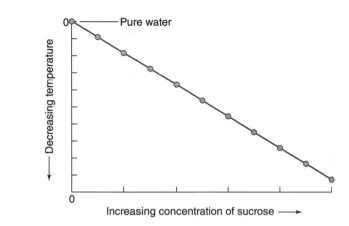

Figure 3–11. Temperature vs. concentration of sucrose

# CHAPTER FOUR

# Gas Laws

# GAS LAWS

In 1643, Evangelista Torricelli invented the first mercury barometer. A few years later, Robert Boyle discovered that the barometer could be used to predict weather. Observations that the volume of air in a sealed tube depends on the air pressure pushing on the surface of mercury became a starting point for the research that led to the gas laws.

# BOYLE'S LAW

Robert Boyle constructed an air pump to study the effect of pressure on the volume of gases. Boyle discovered that at a constant temperature, the volume of a constant mass of gas decreases as pressure increases and increases as pressure decreases. He also discovered that when a constant mass of gas, under constant temperature, is forced into a smaller volume, the gas will push back with more pressure.

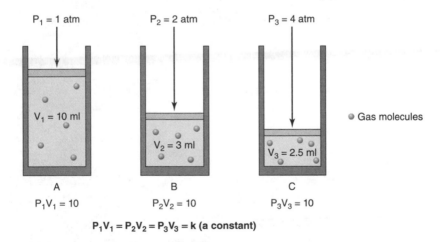

Figure 4–1. Boyle's law cylinder

Pressure and volume have an inverse relationship. When pressure increases, volume decreases. This means that the pressure and volume change in such a way that **the product of pressure and volume is constant**.

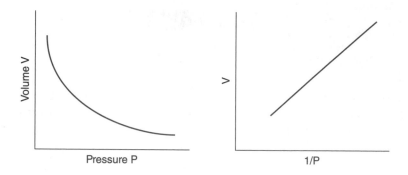

Figure 4–2. Boyle's law graphs

Because pressure and volume are inversely related, the graph of experimental data is a curve. When the data is plotted as volume versus 1/pressure, the curve becomes a straight line. Data may be displayed either way.

## COMMON UNITS OF PRESSURE AND EQUIVALENTS

| Unit of Pressure | mm Hg, (0°C) | torr | atm | kPa |
|---|---|---|---|---|
| millimeters of mercury (mm Hg, 0°C) | 1 | 1 | 0.0013 | 0.1333 |
| torr | 1 | 1 | 0.0013 | 0.1333 |
| atmospheres (atm) | 760 | 760 | 1 | 101.3 |
| kilopascals (kPa) | 7.5 | 7.5 | 0.0098 | 1 |

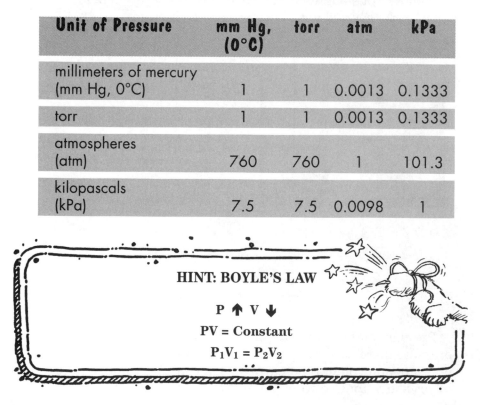

HINT: BOYLE'S LAW

P ↑ V ↓

PV = Constant

$P_1V_1 = P_2V_2$

# BRAIN TICKLERS
### Set #14

***Questions 1 and 2 refer to the diagram below.***

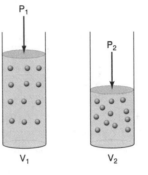

1. If $P_1$ is 1 atmosphere and $V_1$ is 10 liters, what is the constant product of PV?

2. If $V_2$ is 5 liters, what is $P_2$?

***Questions 3 to 5 refer to the graph below.***

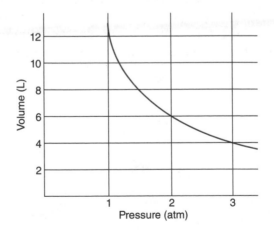

3. According to the graph, how much pressure would be required to compress the gas to a volume of 6 liters?

4. If the gas could be compressed to 2 liters, how much pressure would be needed?

5. If the pressure is less than 1 atmosphere, what should happen to the volume of the gas?

(Answers are on page 91.)

# CHARLES' LAW AND GAY LUSSAC'S LAW

Hot-air balloons became popular in the 1800s. Two French scientists, Jacques Charles and Joseph-Louis Gay-Lussac, conducted carefully controlled experiments to discover the relationship between the temperature and volume of a gas and temperature and pressure of a gas as part of their interest in hot-air ballooning.

## Charles' Law

Jacques Charles discovered that when the mass of gas and pressure are constant, the volume of a gas increases as temperature increases. As a gas cools, the volume decreases. Volume divided by temperature is a constant for a gas.

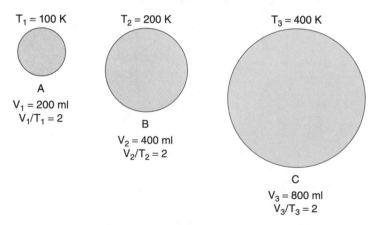

$T_1 = 100$ K

A
$V_1 = 200$ ml
$V_1/T_1 = 2$

$T_2 = 200$ K

B
$V_2 = 400$ ml
$V_2/T_2 = 2$

$T_3 = 400$ K

C
$V_3 = 800$ ml
$V_3/T_3 = 2$

$V_1/T_1 = V_2/T_2 = V_3/T_3 = K$ (a constant)

Figure 4–3. Volume/temperature constant

The change in volume of a gas is directly proportional to the change in temperature. Therefore the graph of volume depends on temperature is a line with a positive slope.

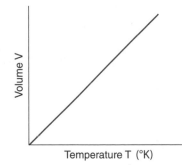

Figure 4–4. Volume vs. pressure

The volume of a gas cannot be a negative number. If the volume of gas could be reduced to zero, the molecular kinetic energy would be at its minimum. The temperature would be absolute zero. Theoretically, absolute zero is –273°C or –459.7°F.

To avoid the mathematical complication of negative numbers, the Kelvin temperature scale is used to measure temperatures below 0°C and in most gas law calculations. On the Kelvin scale, absolute zero is 0°K.

## COMMON REFERENCE TEMPERATURES

| Temperature Scale | Centigrade (°C) | Kelvin (°K) | Fahrenheit (°F) |
|---|---|---|---|
| Absolute temperature | –273 | 0 | –459.7 |
| Freezing point of water | 0 | 273 | 32 |
| Boiling point of water | 100 | 373 | 212 |

**HINT**

To convert centigrade temperatures to Kelvin add 273.
0°C = 273°K

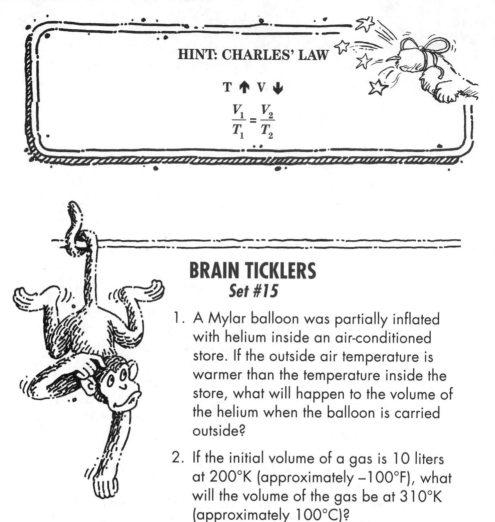

HINT: CHARLES' LAW

T ⬆ V ⬇

$$\frac{V_1}{T_1} = \frac{V_2}{T_2}$$

## BRAIN TICKLERS
### Set #15

1. A Mylar balloon was partially inflated with helium inside an air-conditioned store. If the outside air temperature is warmer than the temperature inside the store, what will happen to the volume of the helium when the balloon is carried outside?

2. If the initial volume of a gas is 10 liters at 200°K (approximately –100°F), what will the volume of the gas be at 310°K (approximately 100°C)?

**Use the graph to answer questions 3 and 4.**

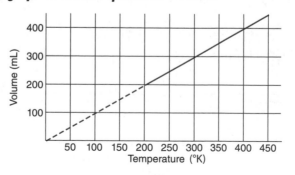

Figure 4–5. Charles' law graph

3. According to the graph, what is the volume of the gas at 300°K?

4. What is the temperature at which the volume of a gas equals zero?

(Answers are on page 91.)

# Gay-Lussac's Law

In 1802, Joseph-Louis Gay-Lussac discovered that when the volume and mass of a gas are held constant, an increase in temperature causes an increase in pressure. Through experimentation, he proved that pressure divided by temperature is constant for a gas.

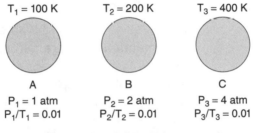

| $T_1 = 100$ K | $T_2 = 200$ K | $T_3 = 400$ K |
|:---:|:---:|:---:|
| A | B | C |
| $P_1 = 1$ atm | $P_2 = 2$ atm | $P_3 = 4$ atm |
| $P_1/T_1 = 0.01$ | $P_2/T_2 = 0.01$ | $P_3/T_3 = 0.01$ |

$$P_1/T_1 = P_2/T_2 = P_3/T_3 = K \text{ (a constant)}$$

Figure 4–6. Gay-Lussac constant

The change in pressure is directly proportional to the change in temperature; therefore the slope of the line of pressure-versus-temperature is positive.

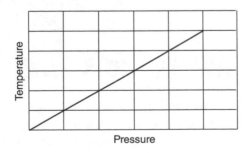

Figure 4–7. Gay-Lussac graph

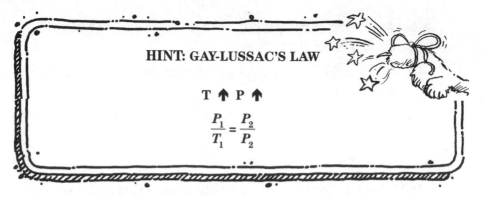

HINT: GAY-LUSSAC'S LAW

T ⬆ P ⬆

$$\frac{P_1}{T_1} = \frac{P_2}{P_2}$$

## BRAIN TICKLERS
### Set #16

1. A bicycle tire was filled with air to the recommended pressure during the day when the temperature was 65°F. Overnight the temperature dropped to 40°F. What will happen to the tire pressure?

2. A playground ball was inflated to 0.5 atmospheres of pressure when the temperature was 288°K (60°F). What is the pressure of the ball when the temperature is 300°K (80°F)? If there were a change, what would you observe?

(Answers are on pages 91–92.)

# AVOGADRO'S LAW

In 1811, Amedeo Avogadro stated that at the same temperature and pressure, equal volumes of all gases contain the same number of molecules. Although Avogadro did not determine the number of molecules, the number was named in his honor. Avogadro's number is $6.02 \times 10^{23}$.

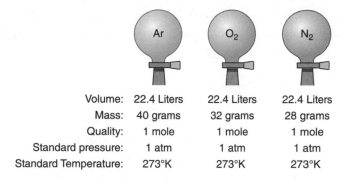

| | | | |
|---|---|---|---|
| Volume: | 22.4 Liters | 22.4 Liters | 22.4 Liters |
| Mass: | 40 grams | 32 grams | 28 grams |
| Quality: | 1 mole | 1 mole | 1 mole |
| Standard pressure: | 1 atm | 1 atm | 1 atm |
| Standard Temperature: | 273°K | 273°K | 273°K |

Figure 4–8. Defining a mole of a gas

At standard temperature (273°K) and pressure (1 atm or 100 kPa) (STP), $6.02 \times 10^{23}$ molecules of a gas occupy a volume of 22.4 liters. Think of the mole (mol) as a measure of the amount of a substance that is a bridge between mass in grams and molecular (molar) mass in amu.

For example, the atomic mass of argon is 40 amu. Argon is a noble gas; therefore the molecular mass is 40 grams. There are 40 grams of argon in 1 mole of argon.

Oxygen and nitrogen occur as diatomic (2 atoms) molecules. The molecular mass of oxygen is 32 amu ($2 \times 16$ amu) or 32 grams in 1 mole of oxygen gas. Nitrogen has a molecular mass of 28 amu ($2 \times 14$ amu); therefore 1 mole of nitrogen is 28 grams.

According to Avogadro's law, if 1 mole of a gas, for example oxygen, occupies 22.4 liters at STP, then 2 moles of a gas will occupy 44.8 liters at STP.

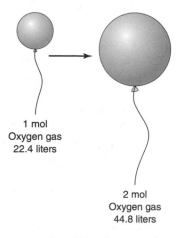

1 mol
Oxygen gas
22.4 liters

2 mol
Oxygen gas
44.8 liters

Figure 4–9. Calculating volume

Avogadro's law can be expressed as an equation, where P is the pressure of the gas, T is the temperature of the gas, and n is the number of moles of gas.

$$\frac{p_1 \cdot V_1}{T_1 \cdot n_1} = \frac{p_2 \cdot V_2}{T_2 \cdot n_2} = constant$$

Figure 4–10. Avogadro's law equation

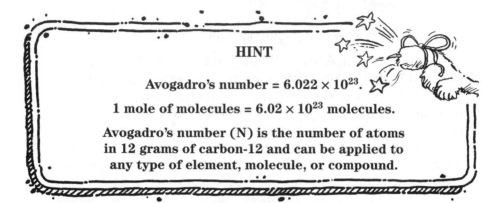

**HINT**

**Avogadro's number = $6.022 \times 10^{23}$.**

**1 mole of molecules = $6.02 \times 10^{23}$ molecules.**

**Avogadro's number (N) is the number of atoms in 12 grams of carbon-12 and can be applied to any type of element, molecule, or compound.**

## BRAIN TICKLERS
### Set #17

1. The atomic mass of the noble gas helium is 4. How many grams of helium are in 1 mole of helium at STP?

2. Hydrogen has an atomic mass of 1 amu and forms a diatomic molecule $H_2$. How many grams of hydrogen are in 1 mole of hydrogen at STP?

3. One mole of neon has a mass of 20 grams and volume of 22.4 liters at STP. What is the volume at STP of 1 mole of chlorine gas if the mass is 70 grams?

4. If 1 mole of nitrogen at STP in a sealed container is heated to 325°K, what will the pressure be?

(Answers are on page 92.)

# IDEAL GAS LAW

Properties of an ideal gas:

- Small particle size.
- No attraction or repulsion interaction between particles.
- Particles spread out.
- No energy is lost when particles collide with each other.
- Collisions with the inside of the container generate pressure.
- Movement (kinetic energy) is proportional to temperature.

In 1834, Emile Clapeyron combined the relationships of pressure, temperature, and volume into a single equation, the ideal gas law. There are no ideal gases, but for most gases, the ideal gas law can be used to predict changes in pressure, volume, and temperature.

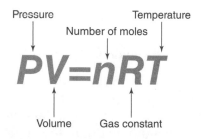

Figure 4–11. Ideal gas law

In the ideal gas law equation, "**n**" represents the number of moles of gas and "**R**" is a gas constant (see hint box).

To analyze the relationships among pressure, volume, and temperature, ignore **n** and **R** and create a proportional expression. Substituting the symbol alpha for the equals sign does this.

If pressure is constant:

$$P_c \; V \; \uparrow \; \alpha \; T \; \uparrow \qquad\qquad P_c \; V \; \downarrow \; \alpha \; T \; \downarrow$$

If volume is constant:

$$P \; \uparrow \; V_c \; \alpha \; T \; \uparrow \qquad\qquad P \; \downarrow \; V_c \; \alpha \; T \; \downarrow$$

If temperature is constant:

$$P \; \downarrow \; V \; \uparrow \; \alpha \; T_c \qquad\qquad P \; \uparrow \; V \; \downarrow \; \alpha \; T_c$$

Note: α means "proportional to."

Figure 4–12. Ideal gas law proportional relationships

Notice that one of the variables must remain constant. Of the two remaining variables, one causes a change in the other. The one that causes the change is the independent variable. The other is the dependent variable.

All balloons contain one mole of gas ($6.02 \times 10^{23}$ molecules)

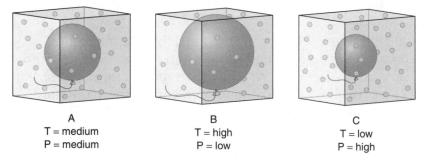

| A | B | C |
|---|---|---|
| T = medium | T = high | T = low |
| P = medium | P = low | P = high |

Figure 4–13. Balloon with 1 mole of gas

The balloons show the effect of changes in external temperature and pressure on gas volume. Each balloon contains 1 mole of a gas. When pressure and temperature outside the balloon are both medium, the volume of gas in the balloon is medium. When temperature is high and pressure is low outside the balloon, the volume of gas inside the balloon increases. When the temperature is low and pressure is high outside the balloon, the volume of the gas inside the balloon decreases. No gas was added or lost; therefore the mass or moles of gas was constant.

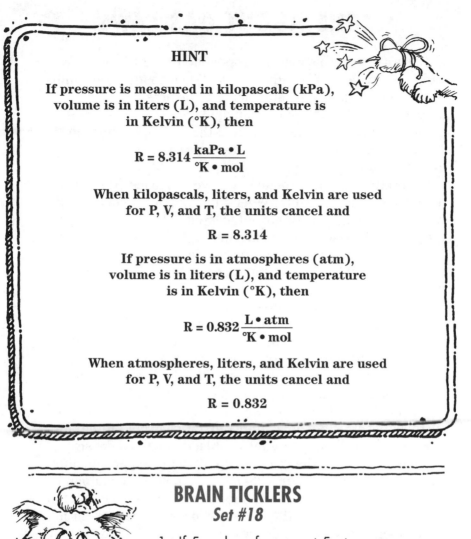

## HINT

If pressure is measured in kilopascals (kPa), volume is in liters (L), and temperature is in Kelvin (°K), then

$$R = 8.314 \frac{kaPa \cdot L}{°K \cdot mol}$$

When kilopascals, liters, and Kelvin are used for P, V, and T, the units cancel and

$$R = 8.314$$

If pressure is in atmospheres (atm), volume is in liters (L), and temperature is in Kelvin (°K), then

$$R = 0.832 \frac{L \cdot atm}{°K \cdot mol}$$

When atmospheres, liters, and Kelvin are used for P, V, and T, the units cancel and

$$R = 0.832$$

## BRAIN TICKLERS
### Set #18

1. If 5 moles of a gas at 5 atm pressure has a volume of 10 liters, what is the temperature of the gas?

2. If 4 moles of a gas in a 60-liter container are at 315°K, what is the pressure in kilopascals?

3. If a gas has a volume of 20 liters, at 200 kPa pressure, and 100°K, how many moles of gas are in the container?

(Answers are on page 92.)

# MORE ABOUT GAS LAWS

In Earth's lower atmosphere (troposphere), air temperature and pressure decrease with altitude. When an airplane climbs to cruising altitude, the cabin needs to be pressurized to prevent passengers from becoming ill or dying. Pressure is created by heated air diverted from the gas engine turbine. This is an example of the pressure increasing as a result of increasing temperature.

Grills attached to a propane tank work because the gas in the tank is under pressure. When the nozzle is opened, the gas will move through the tubing connected to the burners. If the propane tank is exposed to extreme heat, the gas pressure may exceed the ability of the tank to contain it. If the temperature outside the tank is too low, the pressure will be too low for the grill to work efficiently,

Air in Earth's atmosphere collects in areas of high and low pressure because of the unequal heating of Earth's surface. Hot air will rise, creating an area of low pressure. Cold air condenses and sinks, creating an area of high pressure. These pressure differences are what Torricelli observed using his mercury barometer.

The floating of the hot-air balloon through the cooler surrounding air is not an example of a gas law, but rather an example of buoyancy (less dense objects float in a denser medium). However, the gas laws apply to the expansion and contraction of the balloon. When the burner of the balloon is turned on, the air temperature increases and the volume of the balloon expands. When the burner is turned down, the air in the balloon cools and condenses, deflating the balloon.

Setting standards for inflating basketballs and tires, or for filling scuba tanks and aerosol cans, requires knowledge of gas laws. Gas law observations can be made in the classroom using simple setups. Try the brain ticklers to test your overall understanding of the gas laws.

## Caution—Major Mistake Territory!

Air is a mixture of gases. There are additional laws that apply to gas mixtures that are discussed in Chapter 5. When answering the brain ticklers, you may assume that air acts as a single gas.

# BRAIN TICKLERS
## Set #19

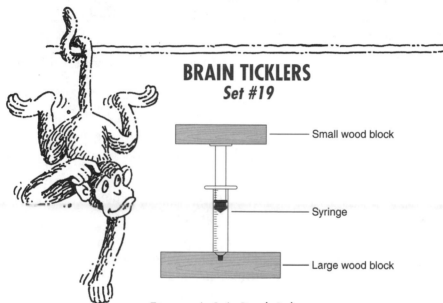

Figure 4–14. Boyle's law apparatus

1. Students placed six 100 g weights on top of the plunger of a sealed syringe filled with 35 ml of air at 22°C (72°F). If the weights are added one at a time, what will happen? Which gas law does this demonstrate? Use evidence to support your answer. Assume that there is no friction between the plunger and the syringe.

2. Students placed a deflated balloon over the top of an empty 2-liter soda bottle. They bent the neck of a desk lamp so that the lightbulb was at the base of the

bottle. The lightbulb was close to the bottle, but did not touch the bottle. The students turned on the lightbulb for 5 minutes. The balloon partially inflated. Which gas law does this demonstrate? Use evidence to support your answer.

3. Students inserted a temperature strip into a 1-liter soda bottle. They screwed on an air pump to the top of the bottle. As they pumped air into the bottle, the temperature increased from 22°C to 30°C and the side of the bottle became more rigid. Use the principles of the ideal gas law to explain what happened. What would you need to know in order to calculate the change in pressure?

(Answers are on page 93.)

## Wrapping Up

- As the pressure on a gas increases, the volume decreases, assuming that the mass and temperature are constant. —Boyle's law

- As the temperature of a gas increases, the volume increases, assuming that mass and pressure are constant.—Charles' law

- As the temperature of a gas increases, the pressure increases, assuming that the mass and volume are constant.—Gay-Lussac's law

- Avogadro's number ($6.02 \times 10^2$) is the number of molecules of a substance in 1 mole of the substance.

- Pressure, temperature, volume, and the moles of a gas are interrelated.—Ideal gas law

- Theoretically the volume of a gas will reach zero at absolute zero.

- The Kelvin temperature scale is based on absolute zero and avoids negative temperatures in calculations.

# BRAIN TICKLERS—THE ANSWERS

## Set #14, pages 77–78

1. $P_1V_1 = 1$ atm $\times$ 10 liters; therefore the constant is 10 atm-liters.

2. 10 atm-liters = $P_2 \times 5$ liters; therefore $P_2 = 2$ atm.
   $$\left( \frac{10 \text{ atm-liters}}{5 \text{ liters}} = 2 \text{ atm} \right)$$

3. The intersection of the y-axis coordinate, 6 liters, with the x-axis coordinate is 2 atmospheres.

4. From the graph it is possible to calculate a constant of 12 atm-liters for this gas. Therefore, 6 atmospheres of pressure would be needed to reduce the volume to 2 liters.

5. If the pressure of the gas was less than 1 atmosphere, the volume should expand to a volume greater than 12 liters. At 0.5 atmospheres, the calculated volume is 24 liters.

## Set #15, pages 80–81

1. Because the temperature is higher outside the store, the volume of helium inside the balloon should increase and more fully inflate the balloon.

2. The volume is 15.5 liters.
   $$\frac{10 \text{ liters}}{200°K} = \frac{15.5 \text{ liters}}{310°K}$$

3. The y-axis coordinate that intersects with the x-axis coordinate of 300°K is 300 ml.

4. The temperature at which a gas has zero volume is absolute zero.

## Set #16, page 82

1. The tire will be softer because when the temperature decreases, the pressure will decrease.

2. The pressure will increase to 5.21 atmospheres. The ball will have slightly more "bounce."

$$\frac{0.5 \text{ atm}}{288°\text{K}} = \frac{0.521 \text{ atm}}{300°\text{K}}$$

## Set #17, pages 84–85

1. At STP, 1 mole of helium has a mass of 4 grams.

2. At STP, 1 mole of hydrogen has a mass of 2 grams.

3. At STP, 1 mole of chlorine gas will have a volume of 22.4 liters.

4. The pressure is 1.19 atm. When the equation is solved, the moles cancel each other, the liters cancel each other, the degrees cancel each other, and the calculation comes down to 325 atm divided by 273, which equals the new pressure.

$$\frac{1 \text{ atm} \bullet 22.4 \text{ liters}}{273°\text{K} \bullet 1 \text{ mol}} = \frac{P_2 \bullet 22.4 \text{ liters}}{325°\text{K} \bullet 1 \text{ mol}}$$

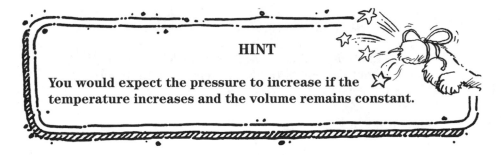

**HINT**

You would expect the pressure to increase if the temperature increases and the volume remains constant.

## Set #18, page 87

1. 12°K

   $$5 \text{ atm} \times 10 \text{ liters} = 5 \text{ mol} \times 0.832 \times \text{T}$$

2. 174.72 kPa

   $$\text{P} \times 60 \text{ liters} = 4 \text{ moles} \times 8.32 \times 315°\text{K}$$

3. 4.8 moles

   $$200 \text{ kPa} \times 20 \text{ liters} = \text{n} \times 8.32 \times 100°\text{K}$$

## Set #19, pages 89–90

1. As the weights are added to the platform on the plunger, they will push down on the air in the syringe. If there is no friction between the plunger and the syringe, the volume will decrease quickly at first. As the air becomes more tightly packed in the syringe, it will push back on the plunger. The volume inside the syringe will continue to decrease, but more slowly until the pressure of the air inside the syringe equals the pressure of the weights on the air. This is an example of Boyle's law because as pressure on a gas increases, the volume of the gas decreases. The temperature would be considered constant because the experiment is at room temperature and any heat generated would be lost to the surrounding air.

2. The lamp transferred heat to the air inside the bottle. The gas molecules in the air gained kinetic energy. They moved upward into the balloon. The balloon allowed the volume of the gas to expand, keeping pressure constant. This demonstrates Charles' law because as the temperature increased, the volume of the gas increased as pressure remained constant.

3. The number of moles of air increased as air was pumped into the bottle. Because the volume of the bottle was constant at 1 liter, the temperature increased as the air molecules experienced increased collisions with one another. The pressure increased as the molecules bounced off the inside of the bottle. Starting and ending volumes and temperatures are known. In order to calculate the change in pressure, the starting and ending number of moles would need to be known.

# CHAPTER FIVE

# Mixtures

# MIXTURES

**Mixtures** are two or more substances that have been combined in such a way that they retain their individual chemical characteristics. Substances in a mixture can usually be separated by ordinary mechanical means using physical properties such as magnetism, density, mass, and particle size. Processes such as boiling, evaporation, and condensation can also be used to separate substances in a mixture.

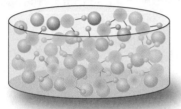 

Homogeneous mixture          Heterogeneous mixture

Figure 5–1. Homogeneous vs. heterogeneous

If the substances are so well mixed they appear to be a single substance, they are said to be **homogeneous**. If the substances appear to be separate components, they form a **heterogeneous** mixture.

**Solutions** are homogeneous mixtures. In solutions the substance that is in the highest concentration is the **solvent**. The minority substances are **solutes**. The solvent **dissolves** the solutes.

When a solid or gas substance dissolves, it is said to be **soluble**. If a solid or gas substance does not dissolve, it is said to be **insoluble**.

If two liquids completely mix into a solution, they are **miscible**. If two liquids do not mix completely into a solution, they are **immiscible**.

## EXAMPLES OF MIXTURES

| Solvent | Solute | Example | Appearance |
|---------|--------|---------|------------|
| Gas | Gas | Nitrogen, oxygen, argon, and other gases in dry air | Homogeneous |

*(Table continues on next page.)*

| Solvent | Solute | Example | Appearance |
|---------|--------|---------|------------|
| Gas | Liquid | Fog—liquid water in gases of air | Homogeneous |
| Gas | Solid | Microscopic particulate matter mixed in air | Homogeneous |
| Liquid | Gas | Oxygen dissolved in water | Homogeneous |
| Liquid | Liquid | Vinegar and water<br>Oil and water | Homogeneous<br>Heterogeneous |
| Liquid | Solid | Salt mixed in water<br>Pepper mixed in oil | Homogeneous<br>Heterogeneous |
| Solid | Gas | Oxygen trapped in ice | Heterogeneous |
| Solid | Solid | Particles of rocks mixed as sand<br>Metal alloys such as bronze (copper and tin) | Heterogeneous<br><br>Homogeneous |
| Solid | Liquid | Silver and mercury in dental fillings | Homogeneous |

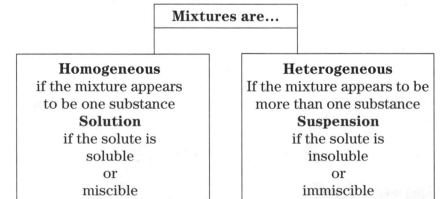

| **Mixtures are…** | |
|---|---|
| **Homogeneous**<br>if the mixture appears<br>to be one substance<br>**Solution**<br>if the solute is<br>soluble<br>or<br>miscible | **Heterogeneous**<br>If the mixture appears to be<br>more than one substance<br>**Suspension**<br>if the solute is<br>insoluble<br>or<br>immiscible |

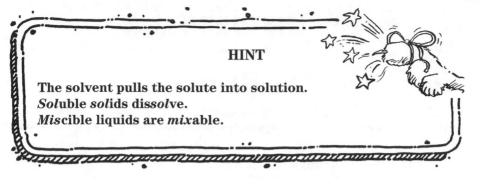

**HINT**

The solvent pulls the solute into solution.
*Sol*uble *sol*ids dis*solv*e.
*Mis*cible liquids are *mix*able.

# BRAIN TICKLERS
## Set #20

1. When salt is mixed with water, why is water the solvent?

2. When water is mixed with gases in air as fog, why is water the solute?

3. Is sugar water a homogeneous or heterogeneous mixture? Explain.

4. Cola soda Is a mIxture of water, sugar, coloring, flavoring, and carbon dioxide. Which best describes the mixture in order of ingredient concentration?
   a. heterogeneous liquid-solid-gas
   b. homogeneous liquid-solid-gas
   c. heterogeneous gas-liquid-solid
   d. homogeneous gas-liquid-solid

5. When pepper is mixed with oil, it becomes suspended in the oil but is visible. Which statement is true?
   a. Pepper is soluble.
   b. Pepper is a solvent.
   c. Pepper is miscible.
   d. Pepper is insoluble.

(Answers are on page 128.)

# GAS MIXTURE LAWS

## Dalton's Law of Partial Pressures

According to **Dalton's law of partial pressures**, each gas in a gas-gas mixture exerts an individual pressure called a partial pressure. The total pressure of the mixture of gases is the sum of the individual pressures.

$$P_{total} = P_1 + P_2 + P_3 \ldots$$

The trailing dots in the equation indicate that there could be more than three partial pressures if there are more than three gases in the mixture.

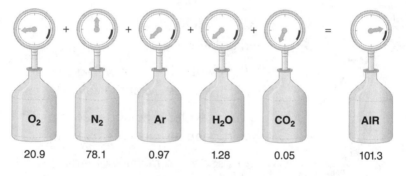

Figure 5–2. Dalton's Law of partial pressures

Air at sea level, a temperature of 0°C (273°K), and 50% humidity exerts a pressure of 101.3 kPa. The total pressure is the sum of the partial pressures exerted by the major gases in the air sample.

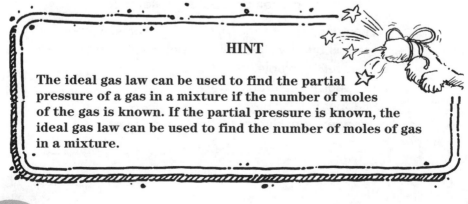

### HINT

The ideal gas law can be used to find the partial pressure of a gas in a mixture if the number of moles of the gas is known. If the partial pressure is known, the ideal gas law can be used to find the number of moles of gas in a mixture.

# Henry's Law

According to **Henry's law** the solubility of a gas in a liquid depends on temperature, the partial pressure of the gas over the liquid, the nature of the liquid, and the nature of the gas.

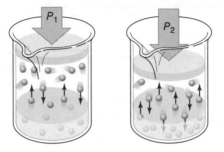

Figure 5–3. Henry's law

In the diagram $P_2$ is greater than $P_1$. The greater pressure above the liquid forces more molecules of gas to remain in solution in the liquid.

When carbon dioxide is bubbled into water to make seltzer, the carbon dioxide gas is held in solution by the pressure of the gas between the liquid surface and the bottle cap. When the bottle is opened, the pressure is released and the carbon dioxide gas begins to bubble up and effervesce (fizz).

When warm seltzer is opened, the carbon dioxide in the water rushes out, taking some of the water with it. Cold seltzer is still bubbly, but less likely to spray out like a fountain. The kinetic energy of the gas is lower at colder temperatures, causing it to stay in the liquid.

When sugar is added to water to make soda, the nature of the liquid changes. Carbon dioxide is more likely to stay in the sugar-water solution than in pure water.

Henry's law is important to manufacturers of carbonated beverages and products that are packaged in high-pressure spray cans. Knowledge of Henry's law avoids overcarbonation that would create excessive pressure or undercarbonation that would affect the flavor of the product. Spray cans depend on the correct amount of propellant gas so that the aerosol is consistent from the first to the last use.

Scuba divers are subject to increases in pressure as they descend into the ocean. If a diver rises to the surface too quickly, nitrogen bubbles form in the diver's blood. Knowledge of Henry's law can help a diver avoid the life-threatening condition called the bends or decompression sickness.

A high-pressure chamber called a hyperbaric chamber was invented to treat decompression sickness. It is also used to raise the partial pressure of oxygen in the blood to treat a variety of disorders, injuries, and illnesses.

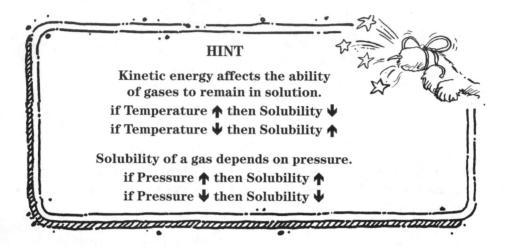

### HINT

**Kinetic energy affects the ability of gases to remain in solution.**
**if Temperature ↑ then Solubility ↓**
**if Temperature ↓ then Solubility ↑**

**Solubility of a gas depends on pressure.**
**if Pressure ↑ then Solubility ↑**
**if Pressure ↓ then Solubility ↓**

## BRAIN TICKLERS
### Set #21

1. Carbon dioxide was collected at 20°C from a chemical reaction by displacement of water from a test tube inverted into a water bath. Some water vapor mixed with the carbon dioxide in the test tube. If the total pressure inside the test tube is 1 atm and the vapor pressure of the water is 0.023 atm, what is the pressure of the carbon dioxide?

2. What will happen to the concentration of dissolved oxygen in water if the water temperature increases?

3. Kal opened a cold 240-ml bottle of cola and drank half of the soda. He replaced the cap and tossed the bottle into his backpack. Later in the day, he opened the cap and heard a whooshing noise. The soda was flat. Explain what happened.

4. When a cold 2-liter soda bottle is first opened, the soda is fizzy. After the leftover soda sits in the refrigerator for a few days, it goes flat even though the cap is screwed on tightly. A company is marketing a small reusable manual air pump that replaces the soda bottle cap. Explain why adding air to the space above the soda keeps the soda from going flat.

5. If the total air pressure at 20°C, at sea level, and 50% humidity is 101.3 kPa and the partial pressure of water vapor in the air sample is 1.28 kPa, what would the air pressure be at 0% humidity?

(Answers are on pages 128–129.)

# SEPARATING MIXTURES

## Magnetism

A magnet can be used to separate a mixture of iron filings and sand. The magnet will attract the iron filings, leaving the sand behind.

## Sifting

A sieve can be used to separate solids of different particle sizes, for example, sand and gravel. The smaller sand particles pass through the sieve, leaving behind the gravel.

# Filtration

Figure 5–4. Filtration

Water can be separated from soil particles by filtration. The soil particles remain on the filter paper lining the funnel while the water collects in the beaker.

## Evaporation and Recrystallization

Salt can be recovered from a salt and water mixture by evaporating the water. Salt crystals will form on the side of the container by recrystallization.

## Distillation

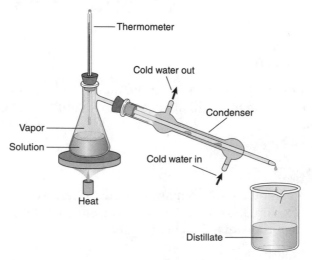

Figure 5–5. Distillation

Two liquids can be separated from each other by distillation if one has a lower boiling point than the other. The first liquid to vaporize flows out through a condenser that is cooled by an outer jacket of water. This causes the vapor to condense back into a liquid and flow out into a container. This process can also be used to recover liquids from a liquid-solid mixture.

# Agitation

Shaking a bottle of soda will separate the carbon dioxide gas from the liquid. After the foam settles, the carbon dioxide will be in the space above the liquid. When the cap is removed, the gas will escape from the bottle. The soda will be less fizzy, or flat.

# Condensation

Cooling a mixture of gases will cause the gases to separate because the gases liquefy at different temperatures. Condensation can also be used to recover liquids that have vaporized.

# Relative Density

When a mixture containing particles suspended in a fluid is allowed to stand undisturbed, the denser particles will settle to the bottom of the fluid and the less dense particles will float to the surface. Liquids of different densities that form a heterogeneous mixture will separate into layers based on relative densities.

For example, in Italian dressing, the oil will float to the surface of the water-vinegar layer and the herbs and spices will sink to the bottom of the water-vinegar layer.

# Paper Chromatography

Paper chromatography can be used to identify substances in a complex solution. A drop of the liquid is placed near one end of a strip of chromatography paper (usually filter paper). The strip of paper is suspended in a solvent so that the drop of liquid is above the solvent. The chromatography tube or chamber is sealed to control atmospheric pressure effects on the solvent vapor pressure.

   The paper is called a stationary medium because it does not move. The solvent is called the mobile medium because it moves up the paper because of capillary pressure on the paper fibers.

   Solutes will migrate (move) up the filter paper with the solvent moving a distance dependent on the solutes' attraction to the paper and to the solvent. The more soluble a substance is, the more distance it moves with the solvent. The diagram shows that a liquid mixture contained two solutes. A reference strip of known substances could be used to identify the two solutes.

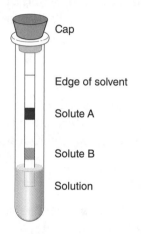

Figure 5–6. Paper chromatography

# BRAIN TICKLERS
## Set #22

### Procedure for separating a mixture of gravel, sand, and salt

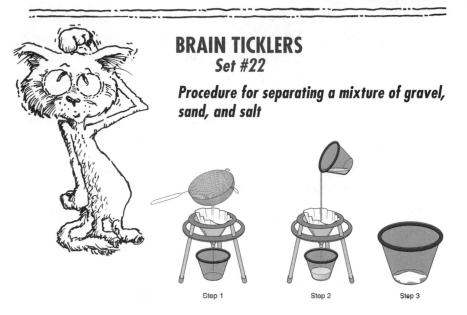

Step 1          Step 2          Step 3

1. In step 1 of the procedure shown in the diagram, a student poured the gravel, sand, and salt mixture through a sieve. The gravel remained in the sieve and the sand and salt fell into the funnel that was lined with filter paper. What physical property did the student use to separate out the gravel?

2. In step 2 of the procedure, the student poured water over the mixture of sand and salt. Sand remained in the filter paper and salt water collected in the cup. Why did the water separate the sand and salt?

3. In the final step, the student recovered salt by evaporating off the water. Evaporation of water from a salt-water mixture recovers salt, but the water is lost as vapor. Describe a procedure that could recover both the salt and the water.

4. A mixture of ethanol and water was separated by distillation. The boiling point of water is 100°C and the boiling point of ethanol is 78.4°C. Which liquid was collected from the condensing tube?

5. A drop of a black liquid was placed on a strip of filter paper. The filter paper was suspended in a solvent. After 15 minutes, the following colors separated into regions from bottom to top: red, blue, brown, black. Which of the following is a valid conclusion relative to solubility?
   a. The red dye molecule is the least soluble.
   b. The blue dye molecule is the least soluble.
   c. The brown dye molecule is the least soluble.
   d. The black dye molecule is the least soluble.

(Answers are on page 129.)

# SOLUTIONS, COLLOIDS, AND SUSPENSIONS

Mixtures involving a **fluid solvent** (gas or liquid) can be classified as solutions, colloids, or suspensions based on particle size.

## Solutions

A **solution** is a homogeneous mixture that forms when the particle size of the solute is less than $10^{-7}$ cm (e.g., salt water). In a solution, the solvent molecules surround the solute particles with the solute particles, filling in spaces between solvent molecules. As the result of thorough mixing, the substances in a solution do not settle or separate over time. Solutions may have color, but they are clear or transparent to light, especially when diluted.

In a perfect solution, a solid solute and solvent are so well mixed that there is no volume change when the solute is added to the solvent. Some combinations of solvents and solutes will actually have a lower volume in solution than the sum of the individual volumes. This occurs most frequently when water is the solvent and the solute molecules are small enough to fit into spaces between the water molecules.

Although the volume may be less than the sum of the component volumes, the mass of the solution is the sum of the mass of the solvent and solute.

## Colloids

**Colloids** are mixtures of solute particle size greater than $10^{-7}$ cm but less than $10^{-5}$ cm (e.g., latex paint, fog). Colloids tend to remain well mixed over time and do not separate into layers on standing. Unlike solutions that are transparent, colloids are cloudy and translucent to opaque.

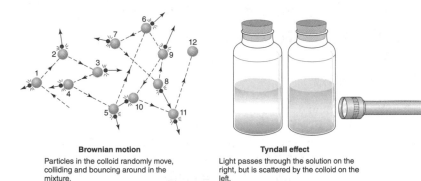

**Brownian motion**
Particles in the colloid randomly move, colliding and bouncing around in the mixture.

**Tyndall effect**
Light passes through the solution on the right, but is scattered by the colloid on the left.

Figure 5–7. Brownian motion and Tyndall effect

When a light beam passes through a **translucent** colloid, for example, fog, the particles are in motion and are visible. The motion is **Brownian movement** (also called Brownian motion) caused by collisions between molecules of the solvent and particles of the solute. The light scattering is the **Tyndall effect**.

Opaque colloids, such as latex paint, do not exhibit the Tyndall effect unless they are diluted with a solvent. A small drop of latex paint in a glass container of water will appear cloudy and exhibit the Tyndall effect.

There are several types of colloids that are classified based on a variety of physical characteristics. Some colloids are more stable than others, but all colloids share the common characteristics of intermediate particle size and homogeneous appearance. Colloids are common natural as well as human-made materials.

| Types of Colloids | Examples |
|---|---|
| Aerosols | Aerosol sprays, smog, natural fog, clouds |
| Solid aerosols | Smoke, dust |
| Foam | Shaving lather, whipped cream |
| Emulsions | Mayonnaise, hand and body lotions, milk |
| Sols | Paint, ink, detergents, latex |
| Solid foams | Marshmallow, Styrofoam |
| Gels | Peanut butter, jelly, gelatin |
| Solid sols | Pearl, opal, some metal alloys |

## Suspensions

**Suspensions** are heterogeneous mixtures with a solute particle size greater than $10^{-5}$ cm (e.g., blood cells in blood serum, sand in water, and flour in water). Suspensions will separate over time by settling. For example, a sample of blood will separate into serum and blood cells.

Substances in a suspension often can be separated by filtration. For example, sand can be filtered out of water.

In translucent suspensions, the particles of the solute are usually visible and appear to be suspended in the solvent. For example, particles of flour reflect light when suspended in water.

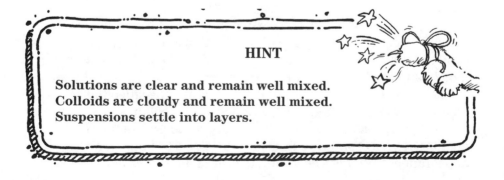

**HINT**

Solutions are clear and remain well mixed.
Colloids are cloudy and remain well mixed.
Suspensions settle into layers.

# BRAIN TICKLERS
### Set #23

1. A mixture of sugar and water is clear and colorless. A beam of light passing through the mixture is not scattered. Is sugar water a solution, colloid, or suspension?

2. Cornstarch has an average particle size of $4 \times 10^{-5}$ cm. In water at room temperature, would cornstarch be likely to form a solution, colloid, or suspension? Explain your answer.

3. When 50 ml of ethanol (alcohol) is mixed with 50 ml of ethanol, the final volume is 100 ml. When 50 ml of water is mixed with 50 ml of water, the final volume is 100 ml. When 50 ml of water is mixed with 50 ml of ethanol, the final volume is 96 ml. Why is the volume of the water-ethanol solution less than 100 ml? (Hint: The ethanol molecule is smaller than the water molecule.)

4. The mass of the 96-ml solution of water and ethanol was equal to the mass of the 50 ml of water added to the mass of the 50 ml of ethanol. Why didn't the mass of the solution decrease?

5. Is hair mousse a solution, colloid, or suspension? Explain.

(Answers are on page 129.)

# SOLUBILITY

When a solid solute mixes completely with a solvent, it is **soluble**. Solid solutes that do not mix with a solvent are **insoluble**.

When a fluid solute mixes completely with a solvent, it is **miscible**. Fluid solutes that do not mix with a solvent are **immiscible**.

Solubility depends on the attraction that occurs between the molecules of the solvent and the molecules of the solute. The attraction is not chemical bonding, but is based on the distribution of charges in a molecule.

Network of Na⁺ and Cl⁻ ions

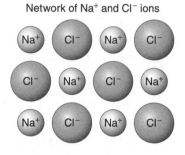

Figure 5–8. Ionic compound

Metal-nonmetal compounds form **ionic bonds** where electrons are transferred from the metal to the nonmetal. In solution, ionic compounds are pulled apart to form ions, or charged particles. The ions will form a solution with a charged solvent. The ionic solution will conduct electricity.

Water, a polar molecule

Ethane, a nonpolar molecule

Figure 5–9. Polar and nonpolar

When a nonmetal element bonds with another nonmetal, the elements share electrons, forming a **covalent bond**. Sometimes electrons are shared unequally, forming areas of positive charge and areas of negative charge. The negative and positive ends of the molecule are called poles and the molecule is said to be **polar**. Covalent molecules that do not have charged poles are called **nonpolar**.

Water is a polar solvent. Salt is an ionic solute. The water molecules surround the salt, pulling the sodium and chloride ions into solution.

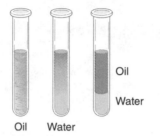

Oil

Water

Oil    Water

Figure 5–10. Oil and water

Corn oil is nonpolar. Water and oil do not form a solution. Corn oil will mix with polar substances such as cornstarch.

Polar solvents form solutions with polar and ionic substances. Most familiar solutions are composed of polar solvents and either polar or ionic solutes. Water is often called the universal solvent because it readily forms solutions with most common ionic solids, polar liquids, and soluble gases.

Nonpolar solvents form solutions with nonpolar substances. Some common nonpolar solvents are fats, oils, and waxes. Some nonpolar solvents that are hazardous and require special handling include gasoline, turpentine, carbon tetrachloride, and benzene.

| Solvent | Solute | Description | | Example |
|---|---|---|---|---|
| Polar liquid | Polar liquid | Miscible | Solution | Water and ethanol |
| Polar liquid | Polar solid | Soluble | Solution | Water and table sugar |
| Polar liquid | Ionic solid | Soluble | Solution | Water and salt |
| Nonpolar liquid | Nonpolar solid | Soluble | Suspension | Corn oil and cornstarch |
| Nonpolar liquid | Polar liquid | Immiscible | Suspension | Corn oil and water |
| Polar liquid | Nonpolar solid | Insoluble | Suspension | Water and cornstarch |

## Rules of Solubility in Water

1. All common compounds of Group I (alkali metals) and ammonium ions are soluble.
2. All nitrates, acetates, and chlorates are soluble.
3. All binary compounds (two different elements) of the group 17 halogens (other than fluorine) with metals are soluble, except those of silver, mercury (I), and lead. Iron halides are soluble in hot water.
4. All sulfates are soluble, except those of barium, strontium, calcium, lead, silver, and mercury (I). Lead, silver, and mercury (I) sulfates are slightly soluble.
5. Carbonates, hydroxides, oxides, silicates, and phosphates are insoluble, unless they are in a compound with a group 1 metal or ammonium ion.
6. Sulfides are insoluble except for calcium, barium, strontium, magnesium, sodium, potassium, and ammonium.

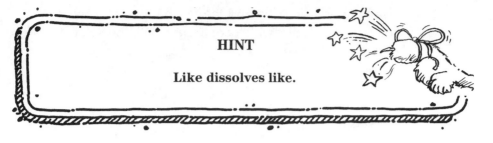

**HINT**

**Like dissolves like.**

# BRAIN TICKLERS
### Set #24

1. Which statement correctly describes the solubility of ethanol in water?
   a. Ethanol is soluble in water.
   b. Ethanol is immiscible in water.
   c. Ethanol is insoluble in water.
   d. Ethanol is miscible in water.

2. Water is a polar molecule. What can you infer about ethanol, a compound of nonmetals?
   a. Ethanol is ionic.
   b. Ethanol is polar.
   c. Ethanol is neutrally charged.
   d. Ethanol is nonpolar.

3. Water will not dissolve oil-based paint from paintbrushes. Mineral spirits will dissolve oil-based paint from paintbrushes. Which best describes the characteristics of mineral spirits?
   a. Polar solvent
   b. Polar solute
   c. Nonpolar solvent
   d. Nonpolar solute

4. Use the periodic table and the solubility rules for water solutions to predict which compounds will be soluble in water. (OH is hydroxide.)
   a. NaCl
   b. KCl
   c. AgCl
   d. NaOH
   e. Ca(OH)$_s$

(Answers are on page 130.)

# EFFECTS OF SOLUTES ON SOLVENT PHYSICAL PROPERTIES

## Conductivity

Solutions involving ionic solutes will conduct electricity. For example, pure distilled water will not conduct electricity. When table salt or other minerals mix with water, the solution becomes an excellent conductor.

## Boiling Point Elevation

A solution has a higher boiling point than the pure solvent. As the amount of solute increases, the boiling point increases. Adding salt to water causes the water to boil at a temperature above 100°C (212°F).

## Freezing Point Depression

A solution has a lower freezing point than the pure solvent. As the amount of solute increases, the freezing point decreases. This also means that the melting point is lower. Adding salt to water causes the freezing point of water to be lower than 0°C (32°F). Ice would also melt at a lower temperature.

## Reduced Vapor Pressure

As the amount of solid nonvolatile solute in a liquid solvent increases, the vapor pressure above the solution decreases. When salt is added to water, the water evaporates more slowly at normal temperatures.

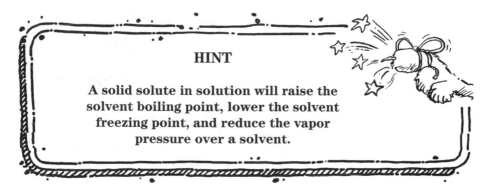

**HINT**

A solid solute in solution will raise the solvent boiling point, lower the solvent freezing point, and reduce the vapor pressure over a solvent.

## BRAIN TICKLERS
### Set #25

1. What would happen to the boiling point of water if sugar were added to the water?

2. Why does soda evaporate more slowly than pure water?

3. Why does antifreeze prevent water in a car radiator from freezing in the winter and boiling in the summer?
   a. The antifreeze raises the freezing point and boiling point of water.
   b. The antifreeze lowers the freezing point and raises the boiling point of water.

    c. The antifreeze raises the freezing point and lowers the boiling point of water.

    d. The antifreeze lowers the freezing point and boiling point of water.

4. Which of the following statement is NOT true about a solution of salt water?

    a. Salt water will conduct electricity.

    b. The boiling point of salt water will be greater than 100°C.

    c. The freezing point of salt water will be higher than 0°C.

    d. The vapor pressure above salt water will be lower than that above pure water.

(Answers are on page 130.)

# FACTORS THAT AFFECT SOLUBILITY

## Temperature

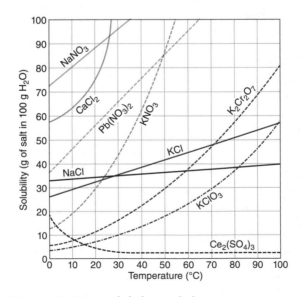

Figure 5–11. Solubility solid vs. temperature

Usually, as the temperature of a solvent increases, the solubility of a solid solute will increase. Increasing the temperature of the solvent increases the kinetic energy of the solvent molecules. This increases the collisions with the solute.

Notice that the effect of a temperature increase depends on the chemical properties of the solute. The solubility of cesium sulfate actually decreases with an increase in water temperature because when cesium sulfate dissolves, it gives off heat. Increasing the temperature of the solvent actually causes the cesium sulfate to drop out of solution.

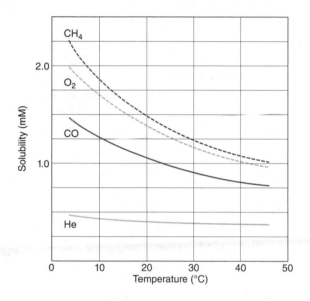

Figure 5–12. Solubility of gases and temperature

As the temperature of a solvent increases, the solubility of a gas solute will decrease. Gases react more quickly to increases in kinetic energy, causing them to escape from the solvent.

An increase in kinetic energy can cause a volatile liquid solute to evaporate from solution. This happens when alcohol and water mix. The escaping alcohol vapor can be observed as arches of liquid on the side of a glass container.

## Surface Area

As the surface area of solute increases, the rate of solubility increases. A cube of sugar will take longer to dissolve than small granules. When the sugar is granulated, the surface area increases, opening new area to action by the solvent molecules.

# Concentration

A solution reaches **saturation** when the solvent can no longer dissolve the solute. A **supersaturated** solution can be created through the use of temperature, pressure, and/or evaporation. By manipulating the conditions, more solute molecules can be forced into solution than would be possible under normal conditions.

# Surfactants

Surfactants such as detergents are chemicals that reduce the surface tension of water. A detergent molecule has one end that is attracted to water—the hydrophilic end. The other end is repulsed by water—the hydrophobic end.

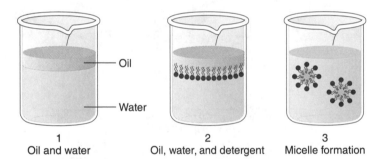

| 1 | 2 | 3 |
| Oil and water | Oil, water, and detergent | Micelle formation |

Figure 5–13. Detergent, oil, and water

1. When oil and water are mixed, they quickly separate into two layers.
2. If a detergent is added to the water and oil, the hydrophilic end of the detergent is attracted to the water layer and the hydrophobic end to the oil layer.
3. After the mixture is agitated, the detergent molecules surround the oil globules, forming micelles.

The detergent holds the oil in the water, forming an emulsion. An **emulsion** is a type of colloid that forms when an immiscible liquid disperses into another liquid by the action of an emulsifier.

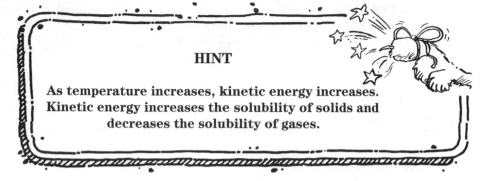

**HINT**

As temperature increases, kinetic energy increases.
Kinetic energy increases the solubility of solids and
decreases the solubility of gases.

## BRAIN TICKLERS
### Set #26

1. Bird feathers are coated with natural oils. During an oil spill, why does crude oil stick to waterbird feathers?

2. Water does not remove crude oil from bird feathers. Why does a solution of water and detergent remove the crude oil?

3. Why does sugar dissolve faster in hot tea than in iced tea?

(Answers are on pages 130–131.)

# CALCULATING CONCENTRATION

Drink mixes have instructions on the product label that specify how much drink powder is needed to make a solution in a variety of common batch sizes. In chemistry, listing the amounts of solute to add to varying amounts of solvent is impractical.

Several mathematical formulas have been developed based on ratios of solute, solvent, or solution. The formulas allow a chemist to mix any volume of solution and to create concentrations needed for specific applications.

# Formulas Based on Mass

**Percent mass** is a way of describing the concentration of the solute in terms of parts per 100. This is used for solutes that are in high concentration in the solution.

$$\text{Percent mass}\left(\frac{\text{mass}}{\text{mass}}\right) = \left(\frac{\text{mass of solute}}{\text{mass of solution}}\right) \times 100$$

**Example:** What is the percentage of sodium chloride in a solution of 20 g of sodium chloride in 2,000 g of solution?

$$\frac{20\text{ g}}{2,000\text{ g}} \times 100 = 1\%$$

When the solution is very dilute, such as gases in liquids or small amounts of particles in gases, it is measured in **parts per thousand** (ppt), **parts per million** (ppm), and **parts per billion** (ppb). These units are often used in environmental chemistry. For example, oxygen dissolved in water is usually measured in milligrams per liter, or parts per million.

$$\text{Parts per thousand} = \left(\frac{\text{mass of solute}}{\text{mass of solution}}\right) \times 1,000$$

$$\text{Parts per million} = \left(\frac{\text{mass of solute}}{\text{mass of solution}}\right) \times 100,000$$

$$\text{Parts per billion} = \left(\frac{\text{mass of solute}}{\text{mass of solution}}\right) \times 1,000,000,000$$

# Formulas Based on Moles

**Molarity** is the number of moles of solute per liter of solution. This is the most commonly used description of concentration for chemical reactions involving acids and bases.

$$\text{Molarity (M)} = \left( \frac{\text{moles of solute}}{\text{liters of solution}} \right)$$

**Example:** What is the molarity of sodium chloride in a solution of 20 g (0.34 moles) of sodium chloride in 2,000 g (2 liters) of solution?

$$\frac{0.34 \text{ moles}}{2 \text{ liters}} = 0.17 \text{ M}$$

**Molality** is usually used for calculations that involve boiling point elevation and freezing point depression.

$$\text{Molality (m)} = \left( \frac{\text{moles of solute}}{\text{kilograms of solvent}} \right)$$

**Example:** What is the molality of sodium chloride in a solution of 20 g (0.34 moles) of sodium chloride in 2,000 g (2 kilograms) of solution?

$$\frac{0.34 \text{ moles}}{2 \text{ kilograms}} = 0.17 \text{ m}$$

**Mole fractions** describe the ratio of solute to solvent or solvent to solute in a solution. Because mole fractions are ratios, there are no units of measure.

$$\text{Mole Fraction}\left( X_{\text{solute}} \right) = \left( \frac{\text{moles of solute}}{\text{moles of solute + moles of solvent}} \right)$$

**Example:** What is the mole fraction of sodium chloride in a solution of 0.34 moles of sodium chloride in 99.66 moles of water?

$$\frac{0.34 \text{ moles}}{0.34 \text{ moles} + 99.66 \text{ moles}} = 0.0034$$

$$\text{Mole Fraction}\left( X_{\text{solvent}} \right) = \left( \frac{\text{moles of solvent}}{\text{moles of solute + moles of solvent}} \right)$$

**Example:** What is the mole fraction of water in a solution of 0.34 moles of sodium chloride in 99.66 moles of water?

$$\frac{99.66 \text{ moles}}{0.34 \text{ moles} + 99.66 \text{ moles}} = 0.9966$$

# Formulas Based on Volume

Concentration may be expressed as a ratio of volumes if both the solute and the solvent are liquids.

$$\text{Volume percent}\left(\frac{v}{v}\right) = \left(\frac{\text{volume of solute}}{\text{volume of solution}}\right) \times 100$$

**Example:** What is the percentage of methanol in a solution of 20 ml of methanol in 2,000 ml of solution?

$$\frac{20\,\text{ml}}{2,000\,\text{ml}} \times 100 = 10\%$$

## HINT

Dilute solutions behave like ideal solutions with respect to volume.
In concentrated solutions, the volume of the solute contributes to solution volume.
When making concentrated solutions, start with the solute and then add the solvent to the desired total volume.

## Caution—Major Mistake Territory!

There are special rules for diluting acids (Chapter 8) and other solutes that release heat in solution. Always dilute by adding high concentration to low concentration.

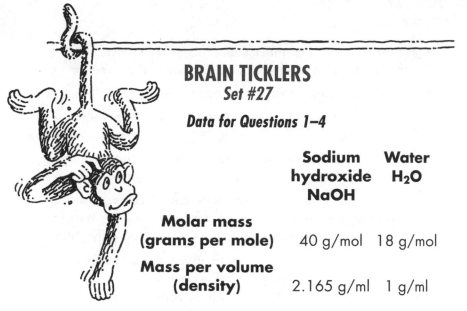

# BRAIN TICKLERS
### Set #27

*Data for Questions 1–4*

|  | Sodium hydroxide NaOH | Water H$_2$O |
|---|---|---|
| **Molar mass (grams per mole)** | 40 g/mol | 18 g/mol |
| **Mass per volume (density)** | 2.165 g/ml | 1 g/ml |

**Questions 1–4:** Use the information in the table to complete the calculations for the following solution: If 50 grams of sodium hydroxide are added to 800 grams of water...

1. What is the percent by mass of sodium hydroxide in the solution?

2. What is the molarity of sodium hydroxide in the solution?

3. What is the molality of sodium hydroxide in the solution?

4. What is the mole fraction of sodium hydroxide in the solution?

5. How many grams of sodium hydroxide would be needed to make 450 ml of a 0.25 molar solution?

6. If 50 ml of ethanol were added to water for a final volume of 225 ml, what is the percentage by volume of ethanol in the solution?

(Answers are on pages 131–132.)

## Wrapping Up

- Mixtures are two or more substances that when combined retain their chemical properties.

- Mixtures that appear to be one substance are homogeneous.

- Mixtures that appear to be more than one substance are heterogeneous.

- The total pressure of the mixture of gases is the sum of the individual pressures.   —Dalton's law of partial pressures

- The solubility of a gas in a liquid depends on temperature, the partial pressure of the gas over the liquid, the nature of the liquid and the nature of the gas.      —Henry's law

- Mixtures can be separated by ordinary means such as magnetism, sifting, filtration, evaporation and recrystallization, distillation, condensation, agitation, relative density, and paper chromatography.

- The three basic types of fluid mixtures are solutions, colloids, and suspensions.

- Solvents dissolve solutes.

- Solubilty depends on solvent temperature, vapor pressure above the solvent, solute surface area, attractions between solvent and solute, and concentration.

- Polar solvents dissolve ionic and polar solutes.

- Nonpolar solvents dissolve nonpolar solutes.

- Polar and nonpolar substances do not mix to form solutions.

- An emulsifier can be used to create a colloid containing polar and nonpolar substances.

- Substances in a colloid exhibit Brownian motion and the Tyndall effect.

- In general, as temperature increases, the solubility of a solid increases.

- In general, as temperature increases, the solubility of a gas decreases.

- Solutes cause freezing point depression, boiling point elevation, and vapor pressure reduction.

- Concentration of solution can be expressed as a percentage of mass, molarity, molality, mole fraction, or percentage of volume.

- Dilute solutions behave as ideal solution with regard to volume.

# BRAIN TICKLERS—THE ANSWERS

## Set #20, page 99

1. Water is the solvent because it is in the highest concentration in the mixture.

2. Water vapor is the solute because it is the lower concentration.

3. When sugar dissolves in water, it appears to be a single substance. Therefore, sugar and water form a homogeneous solution.

4. b. homogeneous liquid-solid-gas
   Cola appears to be one substance. Water is the solvent. The solids (sugar, flavorings, and coloring) and gas (carbon dioxide) are the solutes.

5. d. Pepper is insoluble.
   Pepper does not dissolve in the oil.

## Set #21, pages 102–103

1. 1 atm – 0.023 atm = 0.977 atm
   The partial pressure of carbon dioxide is 0.977 atm.

2. Less oxygen will be dissolved in water as the water temperature increases. Therefore, the concentration of oxygen will decrease as the water temperature increases.

3. The carbon dioxide moved out of the soda into the space above the soda. As the soda warmed up, more carbon dioxide left the liquid and moved into the space, building up pressure. When Kal opened the bottle, he released the pressure and the carbon dioxide left the bottle. The soda lost its fizz when it lost the carbon dioxide.

4. After the first serving of soda, there is a space above the soda. Pumping air into the space creates pressure. If the pressure above the soda increases, then the carbon dioxide

should remain in solution in the soda instead of filling the space above the soda.

5. 101.3 kPa − 1.28 kPa = 100.02 kPa

## Set #22, pages 107–108

1. Particle size

2. Salt is soluble in water. Sand is insoluble in water.

3. The student could boil the water, capture the water vapor, and condense the water vapor back into a liquid. The salt will recrystallize.

4. Ethanol would be collected because it has a lower boiling point.

5. a. The red dye molecule is the least soluble. The red dye has a greater attraction to the paper than to the solvent and drops out of solution first.

## Set #23, page 111

1. Sugar water is a solution because it appears to be a single substance that is clear.

2. At normal room temperatures, based on particle size, cornstarch would probably form a suspension in water.

3. The smaller ethanol molecules fill in the spaces between the water molecules so that the molecules of the solution take up less space than either pure substance.

4. No water or ethanol molecules were lost or gained in the solution; therefore the mass of the solution is the sum of the masses of the solvent and solute.

5. Hair mousse is homogeneous but not transparent; therefore it is not a solution. The solute particles are not visible and do not settle out; therefore it is not a suspension. Hair mousse is a foam colloid.

## Set #24, pages 115–116

1. d. Ethanol is miscible in water.

   *Miscible* is the correct term because both the solvent and the solute are liquids.

2. b. Ethanol is polar.

   Like mixes with like. Water is polar; therefore the nonmetal compound ethanol is polar.

3. c. Nonpolar solvent

   Like mixes with like. Oil is nonpolar; therefore mineral spirits is nonpolar. Mineral spirits is the solvent because it does the dissolving.

4. a. NaCl soluble rule 1, b. KCl soluble rule 1, c. AgCl insoluble rule 3, d. NaOH soluble rule 5, and e. Ca(OH)$_2$ insoluble rule 5

## Set #25, pages 117–118

1. The sugar would raise the boiling point of water.

2. Soda is a solution of water and sugar. As the water evaporates, the sugar becomes more concentrated. This slows evaporation because it lowers the vapor pressure of the water.

3. b. The antifreeze lowers the freezing point and raises the boiling point of water.

4. c. The freezing point of salt water will be higher than 0°C.

   Salt lowers the freezing point of water.

## Set #26, page 121

1. Like mixes with like. The crude oil mixes with the natural oil on the bird's feathers.

2. The detergent makes water wetter by lowering the surface tension. The detergent molecules surround the oil, forming

micelles. The microscopic oil globs can then be washed away with a lot of scrubbing and water.

3. The kinetic energy of the water in the tea is greater, increasing the collisions with the sugar granules. The increased collisions break up the sugar molecules, causing the sugar to go into solution more quickly.

## Set #27, page 125

1. The total mass of the solution is 850 g (the mass of the sodium hydroxide added to the mass of water). Use the percent mass formula:

$$\frac{50\,g}{850\,g} \times 100 = 5.88\%$$

The percent by mass of sodium hydroxide in the solution is 5.88%.

2. There are 1.25 moles of sodium hydroxide in 0.8 liters of water.

$$\frac{1.25\,moles}{0.8\,L} = 0.156\,M$$

The molarity of sodium hydroxide is 1.56 M.

3. There are 1.25 moles of sodium hydroxide in 0.8 kilograms of water.

$$\frac{1.25\,moles}{0.8\,kg} = 0.156\,m$$

The molality of sodium hydroxide is 0.156 m.

4. There are 1.25 moles of sodium hydroxide and 44.44 moles of water.

$$\frac{1.25\,moles}{1.25\,moles + 44.44\,moles} = 0.0027$$

The mole fraction of sodium hydroxide is 0.0027.

5. 4.5 grams of sodium hydroxide in 450 ml would make a 0.25 molar solution.

$$0.1125 \text{ moles} = 0.25 \text{ M} \times .450 \text{ Liters}$$

$$40 \text{ g/mole} \times 0.1125 \text{ moles} = 4.5 \text{ g}$$

6. The percentage by volume of ethanol is 22.2%.

$$\frac{50 \text{ ml}}{225 \text{ ml}} \times 100 = 22.2\%$$

# CHAPTER SIX

# Molecules and Compounds

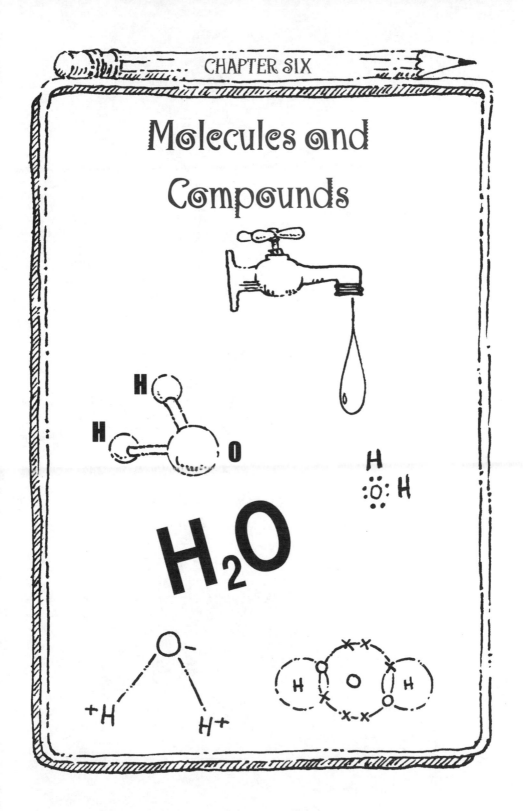

$H_2O$

# MOLECULES AND COMPOUNDS

**Molecules** are composed of at least two atoms that share valence electrons in such a way that both atoms achieve a full valence. **Compounds** are two or more different elements that are connected either by shared electrons or by transferred electrons. The three main types of molecules and compounds are diatomic molecules, molecular compounds, and ionic compounds.

## Diatomic Molecules

If a molecule is composed of two atoms of the **same element**, it is a **diatomic** molecule. Seven elements naturally occur as diatomic molecules: hydrogen, oxygen, nitrogen, fluorine, chlorine, bromine, and iodine. The chemical formulas are $H_2$, $O_2$, $N_2$, $F_2$, $Cl_2$, $Br_2$, and $I_2$.

The **molecular mass** of a diatomic molecule is twice the average atomic mass of the element. (Hint: Average molecular mass of an element is listed on the periodic table of the elements.)

**EXAMPLE:**

Molecular mass of oxygen ($O_2$) = 16 g/mol × 2 = 32 g/mol

## Molecular Compounds

Water, $H_2O$, is a compound in which the elements are joined by shared electrons. Water is an example of a molecular compound because it is both a molecule and a compound. **Molecular compounds** are two or more different **nonmetals**. Glucose ($C_5H_{12}O_6$), water ($H_2O$), and carbon dioxide ($CO_2$) are examples of molecular compounds.

The **molecular mass** of a molecular compound is the sum of the average atomic masses of the elements of the compound.

**EXAMPLE:**

Molecular mass of water ($H_2O$) = 1 g/mol + 1 g/mol + 16 g/mol = 18 g/mol

## Ionic Compounds

Sodium chloride, NaCl, is a compound in which the elements are attracted by transfer of electrons from one element to the other. Sodium chloride is a compound but not a molecule. Sodium chloride (table salt) is an **ionic compound** because it is composed of a **metal and nonmetal**.

The term **formula mass** is used to describe the mass of an ionic compound because technically an ionic compound is not a molecule.

**EXAMPLE:**

Formula mass of sodium chloride (NaCl) = 22.99 g/mol
+ 35.45 g/mol = 58.44 g/mol

## GENERAL PROPERTIES OF TYPES OF COMPOUNDS

| Property | Ionic Compounds | Molecular Compounds | Molecular Crystals |
|---|---|---|---|
| Melting point | high | lower | high |
| Boiling point | high | lower | high |
| Phase at room temperature | solid | solid, liquid, or gas | solid |
| Color | varies | varies | transparent |
| Solubility in water | usually good | only polar | only polar |
| Electrical conductivity as a solid | none | poor | none |
| Electrical conductivity in solution | good | none | none |
| Hardness | hard and brittle | soft | hard |

# Polyatomic Ions

Ionic compounds may include groups of elements called polyatomic ions. **Polyatomic ions** are atoms that are tightly bonded, but as a unit they fail to completely fill the valence of all the partner atoms. Therefore, the unit has a charge and is an ion.

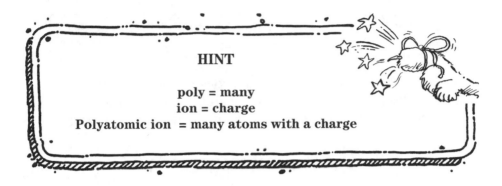

**HINT**

poly = many
ion = charge
Polyatomic ion = many atoms with a charge

Polyatomic ions are sometimes called **molecular ions**, although a few contain metals and are called **metal complexes**. Ions with a positive charge are called **cations**. Ions with a negative charge are called **anions**. Most of the polyatomic ions studied in general chemistry are anions. Ammonium and hydronium are cations that are likely to be included in general chemistry.

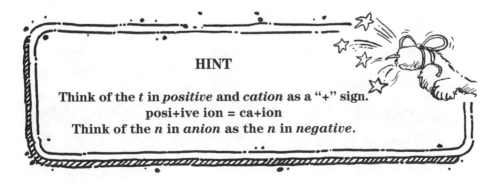

**HINT**

Think of the *t* in *positive* and *cation* as a "+" sign.
posi+ive ion = ca+ion
Think of the *n* in *anion* as the *n* in *negative*.

Polyatomic ions are classified by charge and family. The prefix and suffix provide information about the composition of the ion. For example, the polyatomic ions composed of chlorine and oxygen belong to the chlorine family (see table). They are all anions and have a $-1$ charge. Notice that each has a different number of oxygen atoms.

| Ion Formula | Name | Charge |
|---|---|---|
| $ClO_4^-$ | perchlorate | $-1$ |
| $ClO_3^-$ | chlorate | $-1$ |
| $ClO_2^-$ | chlorite | $-1$ |
| $ClO^-$ | hypochlorite | $-1$ |

An ion that ends in *ate* contains oxygen. The *ite* ending has one less oxygen than *ate*. A *per* prefix indicates one more oxygen than *ate*. The *hypo* prefix would mean one less oxygen than *ite*.

## COMMON POLYATOMIC IONS

| Ion | Formula | Charge | Family |
|---|---|---|---|
| Ammonium | $NH_4^+$ | $+1$ | Cation |
| Hydronium | $H_3O^+$ | $+1$ | Cation |
| Hydroxide | $OH^-$ | $-1$ | Oxygen |
| Nitrite | $NO_2^-$ | $-1$ | Nitrogen |
| Nitrate | $NO_3^-$ | $-1$ | Nitrogen |
| Acetate | $CH_3COO^-$ | $-1$ | Organic |
| Permanganate | $MnO_4^-$ | $-1$ | Metal Complex |
| Hydrogen carbonate (bicarbonate) | $HCO_3^-$ | $-1$ | Carbon |
| Carbonate | $CO_3^{2-}$ | $-2$ | Carbon |
| Peroxide | $O_2^{2-}$ | $-2$ | Oxygen |
| Sulfite | $SO_3^{2-}$ | $-2$ | Sulfur |
| Sulfate | $SO_4^{2-}$ | $-2$ | Sulfur |
| Phosphate | $PO_4^{3-}$ | $-3$ | Phosphorous |

## HINT

Diatomic molecules = two atoms of same nonmetal.
Molecular compounds = two or more
different nonmetals.
Ionic compounds = metals + nonmetals.
Compounds and molecules have a neutral charge.
Ions have a charge.
Polyatomic ions are building blocks
of ionic compounds.

## BRAIN TICKLERS
### Set #28

1. Classify each of the chemicals as a diatomic molecule, molecular compound, ionic compound, or polyatomic ion. Give a reason. (Hint: Use the periodic table of the elements if you are uncertain about whether an element is metal or nonmetal.)

| Chemical | Formula | Description | Reason |
|---|---|---|---|
| Hydrogen | $H_2$ | | |
| Carbon dioxide | $CO_2$ | | |
| Sodium hydroxide | NaOH | | |
| Hydrogen fluoride | HF | | |
| Hydrogen carbonate | $HCO_3^-$ | | |
| Nitrogen | $N_2$ | | |
| Potassium chloride | KCl | | |

2. What is the mass of 1 mole of KCl? (Hint: Use the periodic table of the elements to find the mass of the component elements.) Is this a formula mass or molecular mass? Explain.

3. Analyze the data table for an unknown compound. Is the substance a molecular compound or ionic compound? Use data to support your answer.

| | |
|---|---|
| Phase at room temperature | solid |
| Appearance | white crystals |
| Melting point | 801°C |
| Boiling point | 1,413°C |
| Electrical conductivity in solution | Excellent |

(Answers are on page 160.)

# BONDING

**Bonds** are valence electrons that connect atoms of molecules and compounds to each other. Bonding involves the sharing or transfer of valence electrons. The goal of bonding is a full valence for each member of the molecule or compound. For elements other than hydrogen, bonding results in eight valence electrons. For hydrogen, bonding results in two valence electrons. The three basic types of bonds are ionic, covalent, and metallic.

## Ionic Bonds

**Ionic bonds** form between metal and nonmetal elements. The metal element transfers a valence electron to the nonmetal element. The loss of an electron causes the metal atom to develop a positive charge and become a positive ion. The gain of an electron causes the nonmetal to develop a negative charge and become a negative ion. The two elements remain together because the positive ion is attracted to the negative ion. **Ionic compounds are held together by ionic bonds**.

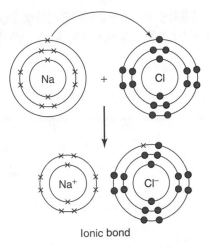

Ionic bond

Figure 6–1. Ionic bond

## Covalent Bonds

**Covalent bonds** form when nonmetals share valence electrons with nonmetals. **Nonpolar** covalent bonds form when atoms of the molecule share electrons equally. **Polar** covalent bonds form when atoms of the molecule do not share electrons equally. If the electrons are not equally shared, poles of negative and positive charges form.

Large carbon compounds such as starches, fats, oils, and waxes tend to be nonpolar. Hydrogen is an example of a nonpolar covalent diatomic molecule. Water is an example of a polar covalent molecular compound that includes an oxygen atom at one end and two hydrogen atoms at the other.

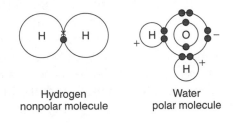

Hydrogen
nonpolar molecule

Water
polar molecule

Figure 6–2. Covalent bond

# PATTERNS FOR PREDICTING POLAR OR NONPOLAR COVALENT BONDS

| Polar covalent bonds | Model | Example | Name |
|---|---|---|---|
| Molecules are linear | AB | CO | carbon monoxide |
| Molecules with a single H | $HA_x$ | HF | hydrogen flouride |
| Molecules with an OH at one end | $A_xOH$ | $C_2H_5OH$ | ethanol |
| Molecules with an O at one end | $O_xA_y$ | $H_2O$ | water |
| Molecules with an N at one end | $N_xA_y$ | $NH_3$ | ammonia |

| Nonpolar covalent bonds | Model | Example | Name |
|---|---|---|---|
| Diatomic molecules of the same element | $A_2$ | $O_2$ | oxygen |
| Carbon forms simple compounds | $C_xA_y$ | $CO_2$ | carbon dioxide |

## Types of Covalent Bonds

The common types of covalent bonds are single, double, and triple bonds. Nonmetal elements that share two valence electrons form a **single bond** between the atoms. A **double bond** involves the sharing of four valence electrons. Six valence electrons are shared in a **triple bond**. A **quadruple bond** or sharing of eight valence electrons occurs in compounds involving some of the middle transition metals, but these are not often encountered in introductory chemistry.

| Bond | Shared electrons | Example | Lewis dot diagram | Orbital diagram |
|------|------------------|---------|-------------------|-----------------|
| Single | 2 | H–H Hydrogen | H:H | |
| Double | 4 | O=O Oxygen | O::O | |
| Triple | 6 | N≡N Nitrogen | N:::N | |

Figure 6–3. Bonds

## Metallic Bonds

When metals are mixed with other metals, they form alloys instead of compounds. **Metallic bonds** between metal atoms are described as a sea of electrons. In metals, the positive nuclei are cations that are fixed while the negative electrons flow around the cations. The free flow of electrons explains why metals are malleable and ductile, have high melting and boiling points, and are good conductors of heat and electricity.

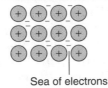

Sea of electrons

Figure 6–4. Metallic bond

## Special Weak Bonds

**Van der Waals** or **London forces** are molecular attractions and repulsions that arise from fluctuations in charges as electrons randomly move around atoms of a molecule, or compound. These weak forces explain why the noble gases, none of which need to gain or lose electrons, will form liquids as they are cooled. Hydrogen bonds are a special category of Van de Waals forces or London dipole forces (*dipole* means "two poles").

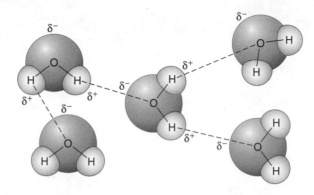

Figure 6–5. Hydrogen bonding

**Hydrogen bonds** form when polar molecules have a positive hydrogen atom that is attracted to a negative atom. Hydrogen bonds are weak but explain cohesion in substances such as water and the boiling points and freezing points of liquids. Two strands of deoxyribonucleic acid are held together in a double helix by hydrogen bonds. Enzymes are another example of large molecules whose shape is determined by hydrogen bonds.

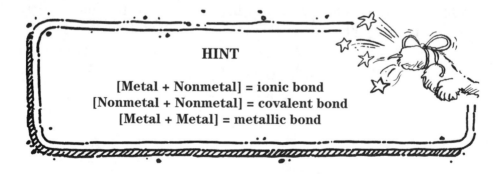

**HINT**

[Metal + Nonmetal] = ionic bond
[Nonmetal + Nonmetal] = covalent bond
[Metal + Metal] = metallic bond

## BRAIN TICKLERS
### Set #29

*Use the periodic table of the elements if you are uncertain about whether an element is metal or nonmetal.*

1. What type of bond will form between copper and tin in the alloy bronze?
   a. ionic
   b. polar covalent
   c. nonpolar covalent
   d. metallic

2. How many electrons are shared by carbon and oxygen in a molecule of carbon monoxide, C:::O? What type of bond is this?

3. Analyze the Lewis dot diagram of methane, $CH_4$. Is this a polar or nonpolar covalent molecule? Explain.

$$\begin{array}{c} H \\ H \colon\! \ddot{C} \colon\! H \\ H \end{array}$$

4. Analyze the Lewis dot diagram of hydrogen fluoride. What holds hydrogen fluoride molecules together in the liquid phase?

$$H \colon\! \ddot{\underset{\cdot\cdot}{F}} \colon$$

(Answers are on pages 160–161.)

# FORMULAS, MODELS, AND DIAGRAMS

## Chemical or Molecular Formula

A **molecular** or **chemical formula** shows the number and types of atoms in a molecule or compound.

The **molecular formula** for glucose is $C_6H_{12}O_6$. The letters indicate the elements of the compound. The subscripts indicate how many atoms of the element are in the compound. The following statements are based on the molecular formula of glucose:

- Glucose contains 6 carbon atoms, 12 hydrogen atoms, and 6 oxygen atoms.

- The **molecular mass** can be calculated as

  $(6 \times 12 \text{ amu}) + (12 \times 1 \text{ amu}) + (6 \times 16 \text{ amu}) = 180 \text{ amu}$

- One mole of glucose has a mass of 180 g/mol.

Although this is useful information for many chemistry applications, the formula does not provide information about the arrangement of the atoms.

The **chemical formula** of an ionic compound provides similar information, except that the term **formula mass** is used instead of *molecular mass* because ionic compounds are not molecules.

## Structural Formulas

A **structural formula** shows the arrangement of atoms in a molecule as well as the number and type of atoms in the molecule. Knowledge of the structure of a molecule is useful for predicting how the molecule will interact with other chemical substances or will behave under changing physical conditions (e.g., temperature and pressure).

**Expanded structural formulas** are important in the study of carbon compounds that have isomers or ring structures. Notice that glucose may be in a straight chain or form two ring isomers in solution. The lines connecting the element symbols show the number of bonds between the atoms of the elements.

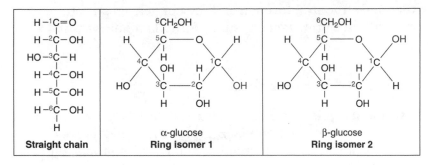

Figure 6–6. Glucose structures

Expanded structural formulas take time to draw and take up a lot of space. A **condensed structural formula** is an abbreviated way of showing structure. The carbons that form the chain are written from left to right along with the elements to which they are attached.

The three isomers of pentane are a good way to see the difference between the information provided by a molecular formula and by structural formulas.

| Molecular formula | Expanded structural formula | Condensed structural formula |
|---|---|---|
| $C_5H_{12}$ | H H H H H<br>H–C–C–C–C–C–H<br>H H H H H | $CH_3CH_2CH_2CH_2CH_3$<br>or<br>$CH_3(CH_2)_3CH_3$ |
| $C_5H_{12}$ | (neopentane structure) | $(CH_3)_4C$<br>or<br>$CH_3C(CH_3)_2CH_3$<br>or<br>$(CH_3)_3CCH_3$ |
| $C_5H_{12}$ | (isopentane structure) | $CH_3CH(CH_3)CH_2CH_3$ |

Figure 6–7. Pentane isomers

All three molecular compounds have the same number of carbon and hydrogen atoms, but the arrangement of the atoms gives each molecule different properties. Look carefully at the

147

expanded structural formula and the condensed structural formula to see how the condensed formula mimics the structural formula.

# Ball and Stick Models

**Ball and stick** models use colored balls to represent atoms and sticks to represent bonds. Ball and stick models are three-dimensional and show the number and types of atoms, the shape of the molecule, and bonding angles between the atoms. These models have applications for physical models and computer simulations.

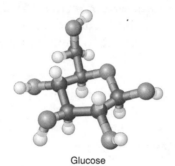

Glucose

Figure 6–8. Ball and stick diagram of glucose

# Lewis Dot Diagrams

It is possible to draw a molecule or compound by combining orbital diagrams for each atom. For small molecules or compounds, this is helpful for understanding how bonds form, but it is not very practical for complex molecules.

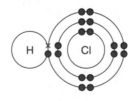

Figure 6–9. HCL diagram

Because bonding involves the valence electrons, the diagram can be reduced to the element symbol and the valence electrons. This is called a Lewis dot diagram.

$$H\!:\!\ddot{\underset{..}{Cl}}\!:$$

## Figure 6-10. HCL Lewis dot diagram

Ionic bonding can also be shown using a Lewis dot diagram. Notice that the ions are often, but not always, displayed in brackets. The brackets indicate that the electrons are not shared, but rather transferred from the metal to the nonmetal. The positive partner has no electrons and a positive superscript. The negative partner has eight electrons and a negative superscript.

$$\left[Na\right]^{+}\left[:\!\ddot{\underset{..}{Cl}}\!:\right]^{-}$$

## Figure 6-11. Ionic Lewis dot diagram

Lewis dot diagrams are shown for elements of period 1-6 and groups 1, 2, and 13-18. The transition metals are not shown because bonding can include electrons from a subshell of a lower orbital than the valence.

| Group | 1 | 2 | | | 13 | 14 | 15 | 16 | 17 | 18 |
|---|---|---|---|---|---|---|---|---|---|---|
| Has* | 1 | 2 | | | 3 | 4 | 5 | 6 | 7 | 8 |
| Needs | -1* | -2 | | | -3 | -4/4 | 3 | 2 | 1 | 0 |
| Charge | +1 | +2 | | | +3 | +4/-4 | -3 | -2 | -1 | 0 |
| Period 1 | H• | | | | | | | | | He: |
| Period 2 | Li• | •Be• | | | •B• | •C• | •N• | :O• | :F• | :Ne: |
| Period 3 | Na• | •Mg• | | | •Al• | •Si• | •P• | :S• | :Cl• | :Ar: |
| Period 4 | K• | •Ca• | | | •Ga• | •Ge• | •As• | :Se• | :Br• | :Kr: |
| Period 5 | Rb• | •Sr• | | | •In• | •Sn• | •Sb• | :Te• | :I• | :Xe: |
| Period 6 | Cs• | •Ba• | | | •Tl• | •Pb• | •Bi• | :Po• | :At• | :Rn: |

☐ Metal ■ Metalloid ☐ Nonmetal

*Hydrogen is a nonmetal that behaves as a metal in some types of compounds.

## Figure 6-12. Periodic table and Lewis diagrams

The top rows show the number of valence electrons each element of a group has, how many electrons the elements would like to lose (–) or gain (+), and the charge an ion has.

Elements of period 2-6 of groups 1, 2, and 13 will have a full valence of eight electrons when the atoms lose the outer electrons. However, the atoms are not stable and become ions. Notice that the ions are positive because as they lose electrons, their neutral atomic charge shifts toward the positive charge of the protons.

Group 14 is a midpoint. The elements of this group have charges of –4 or +4, depending on the bonding partner.

The ions of groups 15, 16, and 17 are negative because they will build a full valence by gaining an electron. The neutrality of the atom shifts toward the negative charge of the electrons.

The elements of group 18 are noble gases. They have full valences and do not need to gain or lose a valence electron. They have a neutral charge and under normal conditions do not form compounds or molecules.

Groups 1 and 17 bond 1:1 to form ionic compounds. Groups 2 and 18 bond 1:1 to form ionic compounds.

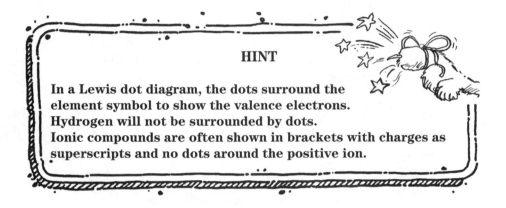

**HINT**

In a Lewis dot diagram, the dots surround the element symbol to show the valence electrons.
Hydrogen will not be surrounded by dots.
Ionic compounds are often shown in brackets with charges as superscripts and no dots around the positive ion.

# BRAIN TICKLERS
### Set #30

*Draw the Lewis dot diagram for each substance.*

1. $Cl_2$

2. $CaO$

3. $HCN$

4. $CCl_4$

5. What information do the structural formulas of ethanoic acid provide compared with the molecular formula?

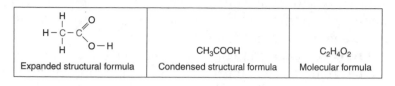

| Expanded structural formula | Condensed structural formula | Molecular formula |
|---|---|---|
| | $CH_3COOH$ | $C_2H_4O_2$ |

6. Analyze the Lewis dot structures for group 1 (alkali metals) and group 17 (halogens) of the periodic table. Why do alkali metals and halogens readily form compounds with each other?

(Answers are on page 161.)

# NAMING COMPOUNDS

In most cases, compounds have a first and last name. The first name is the name of the more positive ion in the compound, and the last name is the more negative ion in the compound.

## Ionic Compounds

Sodium chloride is a good example of how ionic compounds are named. Sodium is the positive ion and retains the name of the element. Chlorine is the negative nonmetal partner. Chlorine becomes chlor*ide*. Notice that the nonmetal element has a suffix *-ide*.

## Compounds with Polyatomic Ions

Compounds involving polyatomic ions follow similar rules, except that the second name is the name of the polyatomic ion, for example, sodium hydroxide. If the compound is composed of two polyatomic ions, the positive ion name goes first, followed by the negative ion name, for example, ammonium carbonate.

# Molecular Compounds

Molecular compounds formed of one type of nonmetal with another type of nonmetal use the name of the most positive nonmetal ion first and the *-ide* form of the more negative ion, for example, hydrogen fluoride.

Prefixes are added to the last name to indicate how many atoms are involved in the molecular compound. The prefixes are also used with the positive nonmetal name if more than one is in the compound. The prefix *mono* is understood if there is only one positive nonmetal. For example, $NO_2$ is nitrogen dioxide, not mononitrogen dioxide. However, $N_2O$ is dinitrogen monoxide.

| Number of atoms | 1 | 2 | 3 | 4 |
|---|---|---|---|---|
| Prefix | mono- | di- | tri- | tetra- |

More elaborate naming rules will be introduced in advanced chemistry courses.

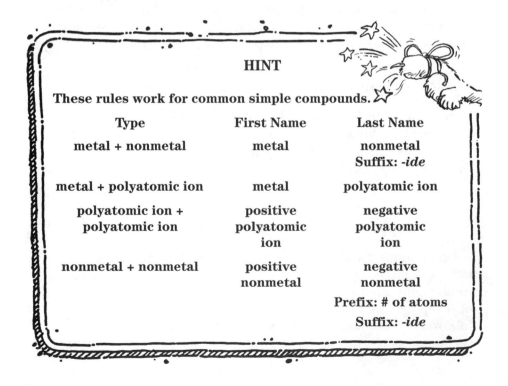

**HINT**

These rules work for common simple compounds.

| Type | First Name | Last Name |
|---|---|---|
| metal + nonmetal | metal | nonmetal<br>Suffix: *-ide* |
| metal + polyatomic ion | metal | polyatomic ion |
| polyatomic ion +<br>polyatomic ion | positive<br>polyatomic<br>ion | negative<br>polyatomic<br>ion |
| nonmetal + nonmetal | positive<br>nonmetal | negative<br>nonmetal<br>Prefix: # of atoms<br>Suffix: *-ide* |

## BRAIN TICKLERS
*Set #31*

*Name the compounds. Use the Periodic Table of the Elements and tables in this chapter if you need help.*

1. $SO_2$.

2. $BaCl$

3. $CaO$

4. $NH_4OH$

5. $CCl_4$

6. $CaCO_3$

7. What is the molecular formula of dihydrogen monoxide?

(Answers are on pages 161–162.)

# CARBON COMPOUNDS

Carbon compounds are found in rocks, foods, solvents, plastics, fuels, fabrics, and all living things on Earth. Thousands of carbon compounds come from crude oil. Living things or organisms depend on four important groups of carbon compounds: carbohydrates, lipids, proteins, and nucleic acids. Because carbon compounds are associated with life processes, carbon chemistry is also known as organic chemistry.

## About Carbon

Carbon forms numerous compounds, because carbon has only two electron shells and four valence electrons. Compounds may be straight chains, branching chains, and rings. Structural formulas are often used to show the arrangement of atoms in complex carbon compounds that have isomers.

Carbon may also form single, double, and triple bonds. Chains containing only single bonds between carbons are **saturated**. Chains containing double bonds between carbons are **unsaturated**.

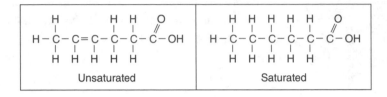

Figure 6–13. Saturated and unsaturated carbon compounds

# Naming Carbon Chains

Carbon chains are named for the number of carbons in the main chain. For example, octane is a compound of carbon and hydrogen that is based on eight carbons. Table 6-1 lists the prefixes that are used for carbon chains from one to ten carbons.

## TABLE 6–1. PREFIXES USED TO NAME CARBON CHAINS

| Number of Carbons in the Chain and Prefix | | | |
|---|---|---|---|
| 1 | meth | 6 | hex |
| 2 | eth | 7 | hepta |
| 3 | but | 8 | oct |
| 4 | prop | 9 | non |
| 5 | pent | 10 | dec |

# Hydrocarbons

Hydrocarbons are molecular compounds that are composed only of carbon and hydrogen. Compounds are named using the prefixes from Table 6–1 and the suffixes (endings) from Table 6–2. There are more naming rules that will be covered in advanced chemistry.

## TABLE 6-2. HYDROCARBON COMPOUNDS

| Type of Compound | Suffix | Functional Group | Example | Name |
|---|---|---|---|---|
| Alkyl group (side branch or chain) | -yl | alkane less 1 terminal hydrogen | 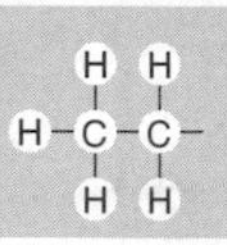 H-C-C- (ethyl structure) | ethyl |
| Alkane $C_nH_{2n+2}$ | -ane | single bonds between carbon atoms | 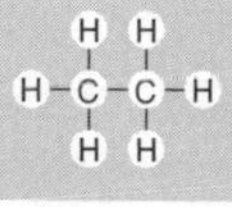 H-C-C-H (ethane structure) | ethane |
| Alkene $C_nH_{2n}$ | -ene | double bond between 2 carbon atoms | 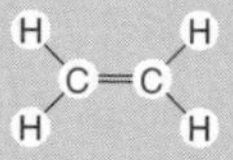 C=C (ethene structure) | ethene |
| Alkyne $C_nH_{2n-1}$ | -yne | triple bond between 2 carbon atoms |  H-C≡C-H | ethyne |
| Cyclic hydrocarbons | cyclo is prefix and -ane, -ene, or -yne as suffix | carbon forms a ring | 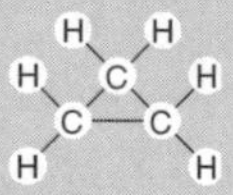 (cyclopropane ring structure) | cyclopropane |

Figure 6-14. Classes of carbon compounds

# Classes of Carbon Compounds

Compounds of carbon, hydrogen, and oxygen are classified according to identifying functional groups. A general formula for each class is shown in Table 6–3. The R can be any number of carbon-hydrogen units. The ending of the formula is the group that determines whether the molecule is an alcohol, aldehyde, ketone, carboxcylic acid, or ester. In some isomers, the functional group may occur in the middle of the molecule.

Names of common examples of each class begin with a prefix from Table 6–1 and end with a suffix from Table 6–3. The relationships among the classes as well as more detailed naming rules will be covered in advanced chemistry courses. For now focus on learning the functional groups and observing patterns.

## TABLE 6–3. COMPOUNDS THAT CONTAIN CARBON, HYDROGEN, AND OXYGEN

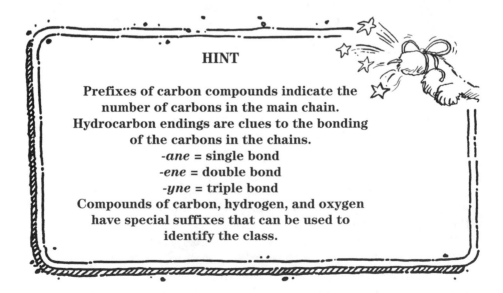

| Class of Compound | Suffix | Functional Group | General Formula | Example | Name |
|---|---|---|---|---|---|
| Alcohol | -ol (alkyl alcohol) | –OH (hydroxyl) | R–OH | | ethanol (ethyl alcohol) |
| Aldehyde | -al | (carbonyl) | R–CHO | | ethanal |
| Ketone | -one | (carbonyl) | R–CO–R' | | propanone |
| Carboxylic acid | -oic acid | (carbonyl) | R–COOH | | ethanoic acid |
| Ester | alkyl–anoate | | R–COO–R' | | methyl ethanoate |

Figure 6–15. Classes of carbon compounds

HINT

Prefixes of carbon compounds indicate the
number of carbons in the main chain.
Hydrocarbon endings are clues to the bonding
of the carbons in the chains.
-*ane* = single bond
-*ene* = double bond
-*yne* = triple bond
Compounds of carbon, hydrogen, and oxygen
have special suffixes that can be used to
identify the class.

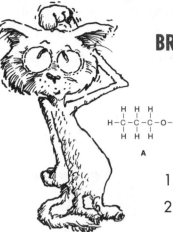

# BRAIN TICKLERS
## Set #32

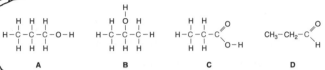

A          B          C          D

1. Which two substances are isomers?

2. Which substance is an aldehyde? Why?

3. What functional group appears on the end of the structural formula of substance C?

4. Another name for fingernail polish remover is acetone. To which class of carbon compounds does it belong?

5. The condensed structural formula of vinegar is $CH_3COOH$. It is sometimes called acetic acid. What is another chemical name for vinegar?

6. Propane is a hydrocarbon fuel used for gas grills. Which of the following is the molecular formula of propane?
   a. $C_2H_6$
   b. $CH_4$
   c. $C_4H_{10}$
   d. $C_3H_8$

7. Is the following an example of a saturated or unsaturated hydrocarbon? Explain.

$$H-\overset{\overset{\displaystyle H}{|}}{C}-\overset{\overset{\displaystyle H}{|}}{\underset{\underset{\displaystyle H}{|}}{C}}-\overset{\overset{\displaystyle H}{|}}{\underset{\underset{\displaystyle H}{|}}{C}}=\overset{}{\underset{\underset{\displaystyle H}{|}}{C}}-\overset{\overset{\displaystyle H}{|}}{\underset{\underset{\displaystyle H}{|}}{C}}-H$$

8. What type of hydrocarbon compound is shown in question 7?

9. The chemical name of rubbing alcohol is isopropanol. What functional group would appear in the structural formula?

(Answers are on page 162.)

## Wrapping Up

- Molecules are two or more atoms that share electrons to achieve a full valence.
- Compounds are atoms of two or more different elements that either share electrons or transfer electrons to achieve a full valence.
- The three main types of molecules and compounds are diatomic molecules, molecular compounds, and ionic compounds.
- Bonds hold atoms of molecules and compounds together.
- Ionic bonds form when a metallic atom transfers an electron to a nonmetal atom.
- Covalent bonds form when atoms of nonmetals share valence electrons.
- Covalent bonds may be single, double, or triple.
- Polar covalent bonds develop a charge because electrons are not shared equally.
- Nonpolar covalent bonds have no charge because electrons are shared equally.
- Metallic bonds are a sea of electrons that give metals the properties of malleability, ductility, good heat and electrical conductivity, and high melting points.
- Polyatomic ions are two or more nonmetallic elements that have a charge after bonding. Metallic complexes are ions of more than one type of element, one of which is a metal.
- Hydrogen bonds are weak.
- Van der Waals or London forces are attractions between polar ends of molecules or ions.

- Molecular and chemical formulas show the numbers and types of atoms in a molecule or compound.

- Structural formulas show the numbers and types of atoms in a molecule or compound as well as the arrangement of the atoms.

- Ball and stick models show the bonding angles and three-dimensional shape of a molecule or compound.

- Lewis dot diagrams show the arrangement of valence electrons around atoms of a molecule or compound.

- Compounds are given chemical names based on the elements that form the compound.

- Organic chemistry is the study of carbon compounds.

- Carbon can form single, double, and triple bonds.

- Carbon atoms form compounds with straight chains, branching chains, and ring structures.

- Structural formulas are useful for understanding the differences among isomers of carbon compounds.

- Common groups are alkyl, alkane, alkene, and cyclical hydrocarbons.

- Common classes of carbon compounds are alcohols, aldehydes, ketones, carboxylic acid, and esters that can be identified by functional groups.

# BRAIN TICKLERS—THE ANSWERS

## Set #28, pages 139–140

| Chemical | Description | Reason |
| --- | --- | --- |
| Hydrogen | Diatomic molecule | Two of the same atoms |
| Carbon dioxide | Molecular compound | Three nonmetals |
| Sodium hydroxide | Ionic compound | Metal with two nonmetals |
| Hydrogen fluoride | Molecular compound | Two different nonmetals |
| Hydrogen carbonate | Polyatomic ion | The substance is made of many atoms and has a charge. |
| Nitrogen | Diatomic molecule | Two of the same atoms |
| Potassium chloride | Ionic compound | Metal with a nonmetal |

2. $\dfrac{39.09 \text{ g}}{\text{mol}} + \dfrac{35.5 \text{ g}}{\text{mol}} = \dfrac{74.59 \text{ g}}{\text{mol}}$

This is a formula mass because KCl is an ionic compound.

3. The substance has high melting point and high boiling point. It is a crystal at room temperature. Because it conducts electricity in solution, it is probably an ionic compound rather than a molecular crystal compound.

## Set #29, page 145

1. d. metallic; copper and tin are both metals

2. Six electrons in a triple covalent bond

3. Methane is nonpolar because the charges are balanced around the chlorine.

4. Hydrogen bonds between the positive hydrogen end and the negative chlorine end will hold the molecules together in the liquid phase.

## Set #30, pages 150–151

1. :C̈l:C̈l:

2. $[Ca]^{+2} [:\ddot{O}:]^{-2}$

3. **H:C:::N:**

4. 
   :C̈l:
:C̈l:C̈l:C̈l:
   :C̈l:

5. Both the structural and molecular formulas show the number and types of atoms in ethanoic acid. The structural formulas both show the arrangement of the atoms. The expanded structural formula shows the types of bonding in the molecule.

6. Alkali metals of group 1 will give away one valence electron to have 8 valence electrons. Halogens of group 17 have 7 valence electrons and need 1 more for a full valence. By forming an ionic compound, both the alkali metal atom and the halogen atom will have 8 electrons in the valence.

## Set #31, page 153

1. $SO_2$, type—nonmetal + nonmetal; name—sulfur dioxide

2. $BaCl$, type—metal + nonmetal; name—barium chloride

3. $CaO$, type—metal + nonmetal; name—calcium oxide

4. $NH_4OH$, type—polyatomic ion + polyatomic ion; name— ammonium hydroxide

5. $CCl_4$, type—nonmetal + nonmetal; name—carbon tetrachloride

6. $CaCO_3$, type—metal + polyatomic ion; name—calcium carbonate

7. Dihydrogen monoxide is $H_2O$.

## Set #32, pages 157–158

1. A and B both have the molecular formula $C_3H_7OH$.

2. D is an aldehyde because it ends in —COH.

3. The carboxyl group, —COOH.

4. Acetone ends in -*one* and is a ketone.

5. Ethanoic acid

6. c. $C_4H_{10}$ ($C_nH_{2n+2}$)

7. Unsaturated because there is a double bond between the third and fourth carbon of the chain.

8. alkene

9. –OH or hydroxyl group

# CHAPTER SEVEN

# Chemical Reactions

# CHEMICAL REACTIONS

A **chemical reaction** occurs when the atoms of starting materials rearrange to form new substances. Matter is not created or destroyed during a chemical reaction. Chemical reactions involve only the valence electrons. The breaking and making of bonds requires energy.

A **chemical change** results in the formation of a new substance or new substances. Energy is needed for a chemical reaction to take place. A chemical change may absorb or release energy usually as heat.

## Signs of Chemical Reaction

Chemical reactions cause changes in both physical and chemical properties of the substances involved in the reactions. One way to tell the difference between a physical change and a chemical change is that a **chemical change is not easily reversed**.

- **Release of light energy**

  Explosive reactions and combustion reactions release energy as light. For example, fireworks explode into a shower of flaming metal salts. This should not be confused with the glow of an incandescent lightbulb caused by resistance to flow of electrons through the bulb's metal filament.

- **Change in temperature**

  The temperature of the products may increase or decrease compared with the starting temperature of the reactants. For example, a chemical ice pack becomes colder after the chemicals inside are mixed or a hot pack becomes warmer when the chemicals react with air. This should not be confused with a change in temperature caused by heating or chilling a substance during a phase change.

- **Formation of a gas**

  Chemical reactions sometimes release a gas that will appear as bubbles or foam. For example, the reaction of baking soda and vinegar releases carbon dioxide gas from the baking soda. This should not be confused with the release of a gas from a gas-liquid mixture such as soda pop.

- **Formation of a precipitate**

  When gas or liquid chemicals are mixed, a solid product may form that settles out of the liquid. For example, when carbon dioxide is bubbled through limewater, the calcium carbonate product is not soluble in water and drops out of solution. This should not be confused with sedimentation from a suspension or a change of phase from liquid to solid.

- **Change in odor**

  A change in odor may indicate that a chemical change has taken place. For example, a burning match releases hydrogen sulfide—the same chemical produced by rotten eggs—or a cake releases a pleasant odor as it bakes. This is different from a change in odor caused by perfume molecules spreading out by diffusion in air.

- **Change in color**

  The product may be a different color from the reactants. For example, when copper reacts with carbon dioxide, water, and oxygen in the air, the copper color changes from reddish orange to green as copper oxide forms. This should not be confused with a mixture such as food coloring in water or color changes that occur as the result of heating or cooling a substance during a phase change (e.g., ice looks white compared with transparent water).

- **Change in pH**

  Certain chemical reactions that involve hydrogen atoms or the polyatomic hydroxide ion can be observed by measuring the change in pH. For example, when lemon juice is added to tea, the pH decreases. Change in pH is discussed in more detail in Chapter 8.

# Controlling Chemical Reactions

Chemical reactions may occur instantly or need help getting started. Changes in reactant starting temperature, surface area, or concentration may affect the rate of chemical reaction. The addition of a catalyst or an inhibitor can speed up or slow down a reaction.

The **rate of reaction** is how long a reaction takes. In introductory chemistry courses, rate is the time elapsed from the start of the reaction to the end of the reaction. In advanced chemistry courses, rate of reaction is expressed in terms of decreasing concentration of reactant per unit of time or increasing concentration of product per unit of time under specified conditions.

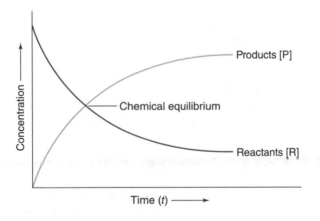

Figure 7–1. Chemical equilibrium

When a chemical reaction begins, the product concentration is zero. The concentration of the reactants will begin to decrease and the concentration of the products will begin to increase.

At some point in the reaction, the concentration of the reactants will equal the concentration of the products. This is called the **chemical equilibrium** point. Some reactions may become stuck at the equilibrium point. Other reactions will continue until all the reactants have been changed into products.

# Changing the Starting Temperature

Raising the starting temperature of a reaction increases the kinetic energy of the system. More kinetic energy increases the interactions between the reactants. In general, an increase in temperature results in an increase in reaction rate. A decrease in starting temperature slows down a reaction.

# Changing the Concentration of Reactants

When there are more particles of reactants, the number of collisions between the particles increases. In general, an increase in reactant concentration increases the reaction rate. A decrease in a reactant concentration slows down the reaction.

# Changing the Surface Area

Grinding a solid reactant into a powder increases the surface area. A greater surface area increases the contact between the reactants. An increase in surface area speeds up a reaction. A decrease in surface area slows down a reaction.

# Catalysts and Inhibitors

Catalysts and inhibitors are sometimes used to control the rate of a chemical reaction. **Catalysts** increase the rate of a reaction without becoming part of the products. **Inhibitors** slow down a reaction without becoming part of the products. **Enzymes** are biological agents that speed up or slow down reactions without becoming part of the products.

## HINT

Chemical reactions result in the formation
of new substances.
Chemical changes are difficult to reverse.
Reactants react.
Products are produced.
Changing the starting kinetic energy
of the reactants changes the rate of
the chemical reaction.

| Increase Reaction Rate | Decrease Reaction Rate |
|---|---|
| Increase Reactant Temperature | Decrease Reactant Temperature |
| Increase Reactant Concentration | Decrease Reactant Concentration |
| Increase Reactant Surface Area | Decrease Reactant Surface Area |
| Add a catalyst | Add an inhibitor |

# BRAIN TICKLERS
## Set #33

A tablet composed of sodium bicarbonate and citric acid was dropped into a beaker of water. The results are shown in diagrams of Experiments A, B, and C.

| Experiment A | Experiment B | Experiment C |
|---|---|---|
| Reaction time 30 sec | Reaction time 120 sec | Reaction time 15 sec |

1. In Experiment A, two tablets were dropped into water at a starting temperature of 22°C. Bubbles were observed. When the bubbling stopped, the temperature of the water had decreased by 1°C. Was this a chemical reaction? Explain.

2. A change in what experimental condition in Experiment B and in Experiment C could explain the change in reaction time? Be specific.

3. Which of the following methods of controlling a chemical reaction involves a **biological** factor?
   a. increasing the starting temperature
   b. decreasing the surface area
   c. adding an enzyme
   d. adding a high concentration of reactants

4. Zinc is a metal that reacts with hydrochloric acid. Which of the following conditions would have the **fastest** reaction rate?
   a. a zinc strip in hydrochloric acid at 20°C
   b. a zinc strip in hydrochloric acid at 30°C
   c. powdered zinc in hydrochloric acid at 20°C
   d. powdered zinc in hydrochloric acid at 30°C

(Answers are on page 187.)

# LAW OF CONSERVATION OF MASS

All of the atoms that enter a chemical reaction end up in the new materials generated by the reaction. During a chemical reaction, bonds between atoms of molecules or compounds break and re-form so that the atoms in the new substances have full valences. Chemical reactions involve only valence electrons.

No new matter is created during a chemical reaction. No matter is lost during a chemical reaction. Therefore, the mass of the reactants is equal to the mass of the products. This is the **law of conservation of mass**. Because no atoms are lost or gained, chemical reactions can be shown by **balanced equations**.

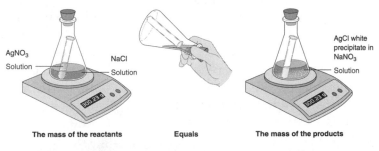

| AgNO₃ Solution | NaCl Solution | AgCl white precipitate in NaNO₃ Solution |
| --- | --- | --- |
| The mass of the reactants | Equals | The mass of the products |

| Chemical reactions can be shown as balanced equations. | | | | | |
| --- | --- | --- | --- | --- | --- |
| $AgNO_3$ | + | NaCl | → | AgCl | + | $NaNO_3$ |
| silver nitrate | and | sodium chloride | yield | silver chloride | and | sodium nitrate |
| Atoms of reactants | | | Atoms of products | | |
| 1 Ag<br>1 Cl<br>1 Na<br>1 N<br>3 O | | | 1 Ag<br>1 Cl<br>1 Na<br>1 N<br>3 O | | |

Figure 7–2. Law of conservation of mass

# Balanced equations

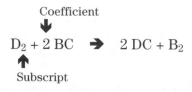

$$A + BC \quad \rightarrow \quad AC + B$$

Reactants     Yield     Products

Figure 7–3. Reaction equation

The starting materials are called **reactants** because they react with one another. The new substances that are formed are called **products** because they are produced by the chemical reaction. The arrow in the equation is read as "yield."

Coefficient
↓
$$D_2 + 2\,BC \quad \rightarrow \quad 2\,DC + B_2$$
↑
Subscript

Figure 7–4. Subscript and coefficient

Atoms combine to form molecules or compounds in specific ratios. The **subscript** shows the number of atoms of an element in the molecule or compound. A **coefficient** is used to show how many moles of a substance are needed for the reaction. If only one mole of a substance is required, no coefficient is used. The one is understood.

According to the model equation, one mole of $D_2$ will react with two moles of BC to yield two moles of DC and one mole of $B_2$. But is the equation balanced?

The subscript goes with the element symbol to its left. There are two atoms of D in the molecule $D_2$. BC is a compound. Two moles of BC are needed; therefore there are two atoms of B and two atoms of C on the reactant side of the equation.

When $D_2$ reacts with two moles of BC, the atoms rearrange to form two moles of the compound DC and a diatomic molecule $B_2$. There are two atoms of D, two atoms of C, and two atoms of B on the product side. The equation is balanced.

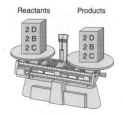

Figure 7–5. Balanced equation

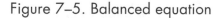

## Analyzing a Chemical Reaction

In 1800, William Nicholson and Johann Ritter passed an electric current through water and captured two gases as the products of the reaction. A simple chemistry class setup of their more elaborate experiment is shown in the diagram.

Figure 7–6. Electrolysis of water

Oxygen gas collected in the test tube labeled A. Hydrogen gas collected in the test tube labeled B. The reaction generated 2 milliliters of hydrogen for each milliliter of oxygen. Hydrogen is a positive ion and oxygen is a negative ion. From these two pieces of information, it is possible to conclude that the molecular formula of water is $H_2O$.

The reactant is $H_2O$. Electricity is the catalyst. The products are hydrogen and oxygen. Both elements form diatomic molecules as a gas at normal room temperature and barometric pressure. The molecular formula for hydrogen is $H_2$ and the molecular formula for oxygen is $O_2$. Using this information, the equation for the reaction is

$$\overset{\text{electricity}}{H_2O \;\rightarrow\; H_2 \uparrow \;+\; O_2 \uparrow}$$

| 2 H | 2 H | 2 O |
| 1 O | | |

Mass of reactant    Mass of products
18 grams          34 grams

Figure 7–7. Decomposition of water

The equation shows a gain of one oxygen atom and with it a gain in mass on the product side. The equation as written **is not balanced**. According to the law of conservation of mass,

mass is neither gained nor lost in a chemical reaction. No atoms should be gained in the chemical reaction.

The equation needs to be balanced. The subscripts cannot be changed because they represent the number of atoms in each molecule. The coefficients can be changed. The balanced equation is

$$\underset{\text{4 H}}{2 H_2O} \xrightarrow{\text{electricity}} \underset{\text{4 H}}{2 H_2} \uparrow + \underset{\text{2 O}}{O_2} \uparrow$$

2 O

| Mass of reactant | Mass of products |
|---|---|
| 36 grams | 36 grams |

Figure 7–8. Decomposition of water

Notice that the subscripts are multiplied by the coefficient to determine the total number of atoms in a molecule. According to the equation, 2 moles of $H_2O$ will yield 2 moles of $H_2$ gas and 1 mole of $O_2$ gas.

The calculation of masses is based on the molecular mass of each substance in the reaction. The gram molecular mass was converted to grams. Under standard conditions, 36 grams of water will produce 4 grams of hydrogen gas and 32 grams of oxygen gas.

Often an upward-pointing arrow is used to indicate that a gas has been produced or (g) will appear next to the molecular formula. Sometimes (l) is used to indicate a liquid and (s) is used for solid. When (aq) appears in an equation, it means that the substances are in a water solution (*aq* stands for "aqueous," from the Latin for "water").

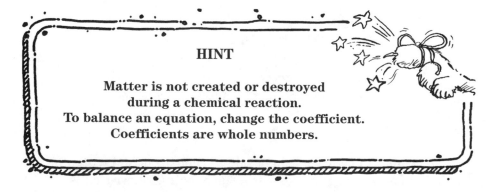

**HINT**

Matter is not created or destroyed
during a chemical reaction.
To balance an equation, change the coefficient.
Coefficients are whole numbers.

# BRAIN TICKLERS
## *Set #34*

1. When 12 grams of carbon (C) react with exactly 32 grams of oxygen (O), carbon dioxide is formed. The mass of the carbon dioxide is
   a. greater than 44 grams
   b. equal to 44 grams
   c. less than 44 grams
   d. not determinable

2. A chemical equation shows
   a. elements and mixtures
   b. hydrogen and oxygen
   c. chemical bonds
   d. reactants and products

3. A chemical equation can be balanced by changing the
   a. reactants
   b. products
   c. coefficients
   d. formulas

**Questions 4 and 5 refer to the following equation**

$$NaOH + HCl \rightarrow NaCl + H_2O$$

4. In this equation, NaOH and HCl are
   a. reactants
   b. coefficients
   c. products
   d. elements

5. Is this a balanced equation? Prove your answer.

6. The subscript 2 in $H_2O$ means that
   a. there are two molecules of hydrogen in water
   b. there are two atoms of oxygen in water
   c. there are two molecules of water
   d. there are two atoms of hydrogen in water

7. Which of the following is a balanced equation?
   a. $Zn + HCl \rightarrow ZnCl_2 + H_2$
   b. $H_2O \rightarrow H_2 + O_2$
   c. $Mg + 2\,HCl \rightarrow MgCl_2 + H_2$
   d. $H_2SO_4 + NaOH \rightarrow Na_2SO_4 + H_2O$

8. The chemical reaction between acetic acid (vinegar) and sodium bicarbonate (baking soda) is shown in the equation below.

$$CH_3COOH + NaHCO_3 \rightarrow CH_3COONa + H_2O + CO_2 \uparrow$$

The sodium bicarbonate was added to the acetic acid in an open beaker. The total starting mass of the acetic acid and sodium bicarbonate was 144.06 grams. The ending mass of the products was 100.06 grams. Explain this apparent violation of the Law of conservation of mass.

(Answers are on pages 187–188.)

# TYPES OF REACTIONS

## Combustion

Combustion is burning. This type of reaction requires oxygen, heat, and something that will burn. The reaction releases heat that may be converted into light energy as a flame or glow.

## Combustion of Hydrogen

Combustion of hydrogen gas produces water. Placing a burning match at the lip of a test tube of pure hydrogen will produce a popping noise and water droplets in the test tube.

$$2\,H_2 + O_2 \;\rightarrow\; 2\,H_2O$$

Figure 7–9. Combustion of hydrogen

## Combustion of Carbon

Combustion of pure carbon produces carbon dioxide. Carbon dioxide extinguishes a flame.

$$C + O_2 \;\rightarrow\; CO_2$$

Figure 7–10. Combustion of carbon

## Combustion of Hydrocarbons

Combustion of hydrocarbons forms carbon dioxide and water. The equations show the combustion of methane (natural gas), octane (a component of gasoline), and ethanol (a fossil fuel alternative). The products of combustion are carbon dioxide and water. When combustion is incomplete, a deadly gas, carbon monoxide is produced.

$$CH_4 + 2\,O_2 \;\rightarrow\; CO_2 + 2\,H_2O$$

Figure 7–11. Combustion of methane

$$2\,C_8H_{18} + 25\,O_2 \;\rightarrow\; 16\,CO_2 + 18\,H_2O$$

Figure 7–12. Combustion of octane

$$C_2H_5O + 3\ O_2 \rightarrow 2\ CO_2 + 3\ H_2O$$

Figure 7–13. Combustion of ethanol

The burning of all hydrocarbon fuels produces carbon dioxide. As a gas, carbon dioxide traps heat. This has become an internationally debated environmental concern about the effect that higher levels of carbon dioxide may have on global average temperatures.

The environmental concern over continued use of fossil fuels is that the carbon in the fuels is carbon that was stored from ancient carbon cycles. The combustion of fossil fuels adds this ancient carbon to the modern carbon cycle, increasing the amount of carbon available for conversion to carbon dioxide.

Current carbon is carbon from today's plants and animals. Ethanol from fermentation of plant materials and methane from decomposition of plant and animal materials are considered fuels that contain current carbon. Although combustion of these fuels releases carbon dioxide into the atmosphere, it is a cycling of current carbon.

# Synthesis

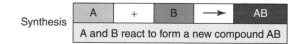

Figure 7–14. Synthesis reaction

*Synthesis* means to make something new by combining two or more materials. A synthesis reaction makes a single product from two or more reactants.

# Decomposition

Figure 7–15. Decomposition reaction

*Decomposition* means to break down something complex into its elements or component parts. Decomposition reactions begin with a single reactant and end with two or more products containing parts of the original reactant.

# Replacement

*Replacement* means that a negative ion or polyatomic ion group moves from one positive partner ion to another. There are two basic replacement reactions: single replacement and double replacement.

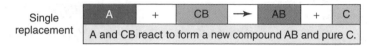

Single
replacement

| A | + | CB | → | AB | + | C |

A and CB react to form a new compound AB and pure C.

Figure 7–16. Single replacement

In a single replacement reaction, one reactant is a metal or a diatomic molecule (e.g., $H_2$) and the other is a molecular or ionic compound. Notice that in the product the negative ion or polyatomic ion of the compound has changed partnership. The products are a new compound and diatomic molecule or metal.

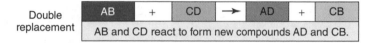

Double
replacement

| AB | + | CD | → | AD | + | CB |

AB and CD react to form new compounds AD and CB.

Figure 7–17. Double replacement reaction

In a double replacement reaction, both reactants are molecular or ionic compounds. Notice that in the product the negative ions or polyatomic ions have changed partners. The products are two new compounds.

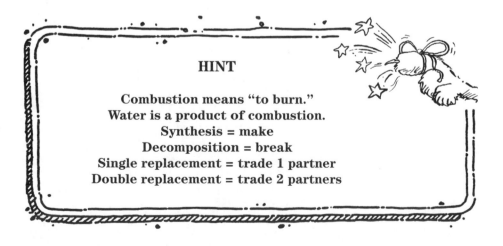

**HINT**

Combustion means "to burn."
Water is a product of combustion.
Synthesis = make
Decomposition = break
Single replacement = trade 1 partner
Double replacement = trade 2 partners

## BRAIN TICKLERS
*Set #35*

*Classify each reaction as combustion, synthesis, decomposition, single replacement, or double replacement.*

| Reaction | Type |
|---|---|
| $2\ Ag + H_1S \rightarrow Ag_2S + H_2$ | |
| $C_6H_{12}O_6 \rightarrow 2\ C_2H_5OH + 2\ CO_2$ | |
| $Ca(OH)_2 + 2\ HCl \rightarrow CaCl_2 + 2\ H_2O$ | |
| $C_3H_8 + O_2 \rightarrow 3\ CO_2 + 4\ H_2O$ | |
| $4\ Fe + 3\ O_2 \rightarrow 2\ Fe_2O_3$ | |
| $Zn + 2\ HCl \rightarrow ZnCl_2 + H_2$ | |
| $HCl + NaOH \rightarrow H_2O + NaCl$ | |

(Answers are on page 188.)

# REACTION ENERGY

Energy is needed to start a chemical reaction. The energy needed to raise the reactants from their starting energy past the reaction equilibrium point is called **activation energy**.

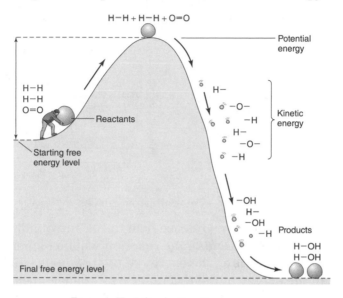

Figure 7–18. Activation energy

Think of the reactants as having the potential to react. Activation energy does the work needed to push the reactants over the edge into action. As the reaction progresses, particles of the reactants move around and collide, converting potential energy to kinetic energy. Because temperature is a measure of average kinetic energy, less kinetic energy will result in a lower temperature and more kinetic energy will result in a higher temperature.

A catalyst works by lowering the "hill" needed to tip reactants into action. Catalysts allow reactions to occur at lower activation energies. This is especially important in biochemistry where the temperature needed for a reaction would destroy the reactants.

A chemical reaction involves the breaking and formation of bonds. Energy is needed to break bonds, and energy is released when bonds form. Energy is measured in joules (J) and often expressed as kilojoules (kJ). Heat is the energy involved in breaking and making bonds.

181

The heat content of a system is called **enthalpy**. Enthalpy takes into account the total potential and kinetic energy of the system. The enthalpy of a reaction at any given point of the reaction would be difficult to measure.

However, all chemical reactions result in a change in energy that can be measured as a change in heat. The symbol ΔH (delta H), for change in enthalpy, is used to show a change in energy.

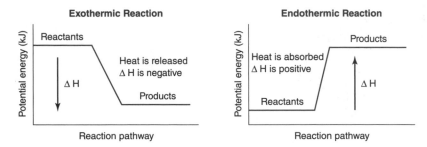

Figure 7–19. Enthalpy curves

**Exothermic reactions** release heat to the surroundings. The temperature at the end of the reaction will be higher than the starting temperature. The energy of the products will be lower than the energy of the reactants. The change in enthalpy is negative. An air-activated chemical hand warmer is an example of an exothermic reaction.

Reactant ➔ Products + Heat Energy
**Reactant Energy** ➔ Product Energy
−ΔH

Figure 7–20. Exothermic model equation

**Endothermic reactions** take in energy from the surroundings. The temperature at the end of the reaction will be lower than the starting temperature. The energy of the products will be higher than the energy of the reactants. The change in enthalpy is positive. A chemical cold pack is an example of an endothermic reaction.

Reactant ➔ Products − Heat Energy
Reactant Energy ➔ **Product Energy**
+ΔH

Figure 7–21. Endothermic reaction model equation

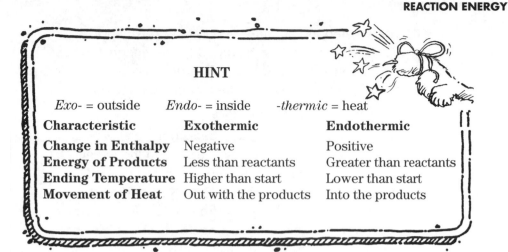

**HINT**

*Exo-* = outside     *Endo-* = inside     *-thermic* = heat

| Characteristic | Exothermic | Endothermic |
|---|---|---|
| **Change in Enthalpy** | Negative | Positive |
| **Energy of Products** | Less than reactants | Greater than reactants |
| **Ending Temperature** | Higher than start | Lower than start |
| **Movement of Heat** | Out with the products | Into the products |

# BRAIN TICKLERS
## Set #36

1. The energy needed to start a reaction is called
   a. heat energy
   b. potential energy
   c. endothermic energy
   d. activation energy

2. If more energy is released by the formation of bonds than is required to break the bonds, a reaction will be
   a. at equilibrium
   b. endothermic
   c. exothermic
   d. static

3. All chemical reactions result in a change in
   a. phase
   b. enthalpy
   c. concentration
   d. mass

4. The starting temperature of a beaker of vinegar was 22°C. Baking soda was added to the vinegar. When the bubbling stopped, the temperature of the liquid in the beaker was 17°C. Is the reaction exothermic or endothermic? Explain your answer.

5. The starting temperature of a beaker of water was 22°C. Yeast, a living organism, was added to the water. No change in temperature was noted. Peroxide was added to the beaker of water and yeast. The liquid began to foam, releasing a strong odor. As the foaming subsided, the temperature was 38°C. Describe what happened in terms of energy transfers. (Hint: Yeast contains the enzyme catalyst peroxidase.)

6. Analyze the equation for the combustion synthesis of liquid water and for the decomposition of liquid water. Which reaction is exothermic and which is exothermic? Support your answer with evidence. Assume standard condition for temperature and pressure.

## Synthesis of Water

Flame

$$2 H_2 + O_2 \rightarrow 2 H_2O$$
$$\Delta H = -285.83 \text{ kJ}$$

## Decomposition of Water

Electricity

$$2 H_2O \rightarrow 2 H_2 + O_2$$
$$\Delta H = 285.83 \text{ kJ}$$

(Answers are on page 189.)

# Wrapping Up

- Chemical reactions result in the formation of a new substance or substances.

- Chemical reactions involve the sharing or transfer of valence electrons.

- Reactants are the chemicals that start the reaction.

- Products are the chemicals that result from the reaction.

- Chemical reactions rates are affected by starting temperature, reactant concentration, reactant surface area, and the presence of a catalyst.

- A catalyst is used to start a reaction, but does not become part of the products.

- Inhibitors slow down or stop a reaction.

- Enzymes are biological catalysts or inhibitors.

- Six signs of a chemical reaction are a change in temperature, a change in color, a change in odor, a change in pH, the formation of a gas, or the formation of a precipitate.

- In a chemical reaction, atoms of reactants are rearranged to form products with no loss or gain of mass.

  —Law of conservation of mass

- Chemical reactions can be shown by an equation that shows the number of moles and chemical composition of each reactant and each product.

- Chemical equations are balanced by changing the coefficients or moles of the reactants or products.

- There are four basic types of chemical reactions: synthesis reactions, decomposition reactions, replacement reactions, and combustion.

- Energy is required to break and make bonds.

- Activation energy is the energy needed to start a reaction.

- Enthalpy is energy released or absorbed during the formation of products by a chemical reaction.

- Reactions that release heat to the surrounding environment are exothermic.

- The enthalpy of an exothermic reaction is lower at the end of the reaction than at the start of the reaction. The change is negative.

- Reactions that absorb heat from the surrounding environment are endothermic.

- The enthalpy of an endothermic reaction is higher at the end of the reaction than at the start of the reaction. The change is positive.

- The change in enthalpy of a reaction can be shown using a graph.

# BRAIN TICKLERS—THE ANSWERS

## Set #33, pages 170–171

1. Experiment A is a chemical reaction. A gas forms and there is a change in temperature. It would be difficult to reverse the reaction.

2. In Experiment B, the starting temperature of the water was colder than in Experiment A. The decrease in temperature decreased the kinetic energy. The reaction was slower. In Experiment C, the starting temperature of the water was warmer than in Experiment A. The increase in temperature increased the kinetic energy. The reaction was faster.

   The diagrams do not support a change in surface area as an answer. The two tablets were not altered to increase surface area.

   The diagrams do not support a change in concentration as an answer. Two tablets added at the same time were constant in all three experiments.

   The diagrams do not support the use of a catalyst or inhibitor. Nothing was added to the water except two tablets.

3. c. adding an enzyme. The other answers are changes in physical conditions.

4. d. powdered zinc in hydrochloric acid at 30°C. This set of conditions. includes increased surface area and increased starting temperature.

## Set #34, pages 175–176

1. b. equal to 44 grams (Law of conservation of mass)

2. d. reactants and products

3. c. coefficients

4. a. reactants

5. The equation is balanced because there is the same number of atoms of each element on both sides of the equation.

| Reactants | Products |
|-----------|----------|
| 1 Na | 1 Na |
| 1 O | 1 O |
| 2 H | 2 H |
| 1 Cl | 1 Cl |

6. d. there are two atoms of hydrogen in water

7. c. $Mg + 2 HCl \rightarrow MgCl_2 + H_2$

8. The upward arrow next to the molecular formula of carbon dioxide shows that carbon dioxide gas was produced. The missing mass is the mass of the carbon dioxide that left the beaker. To prove that the missing mass is the escaped carbon dioxide, the experiment would need to be repeated with the gas trapped in the reaction container or captured by some other means.

## Set #35, page 180

| Reaction | Pattern | Type |
|----------|---------|------|
| $2 Ag + H_1S \rightarrow Ag_2S + H_2$ | $A + CB \rightarrow AB + C$ | Single replacement |
| $C_6H_{12}O_6 \rightarrow 2 C_2H_5OH + 2 CO_2$ | $AB \rightarrow A + B$ | Decomposition |
| $Ca(OH)_2 + 2 HCl \rightarrow CaCl_2 + 2 H_2O$ | $AB + CD \rightarrow AD + CB$ | Double replacement |
| $C_3H_8 + O_2 \rightarrow 3 CO_2 + 4 H_2O$ | hydrocarbon + oxygen $\rightarrow$ carbon dioxide + water | Combustion |
| $4 Fe + 3 O_2 \rightarrow 2 Fe_2O_3$ | $A + B \rightarrow AB$ | Synthesis |
| $Zn + 2 HCl \rightarrow ZnCl_2 + H_2$ | $A + CB \rightarrow AB + C$ | Single replacement |
| $HCl + NaOH \rightarrow H_2O + NaCl$ | $AB + CD \rightarrow AD + CB$ | Double replacement |

## Set #36, pages 183–184

1. d. activation energy

2. c. exothermic

3. b. enthalpy

4. The ending temperature of the reaction was lower than the starting temperature; therefore, the reaction was endothermic. The energy of the products is lower than the energy of the reactants.

5. The peroxidase in the yeast acted as a catalyst. The catalyst lowered the activation energy of the reaction. The reaction released energy in the form of heat, causing the temperature to rise. This is an exothermic reaction.

6. The formation of water by combustion releases energy. The change in enthalpy is negative. The reaction is exothermic. When liquid water is decomposed into hydrogen and oxygen gases, more energy is needed to break the bonds than is released to form new bonds. The change in enthalpy is positive. The reaction is endothermic.

# CHAPTER EIGHT

# Acids, Bases, and Salts

# PROPERTIES OF ACIDS AND BASES

Acids and bases are groups of compounds that react with each other to form a third type of compound salts. The term *acid* is from the Latin *acere*, meaning "sour." Another term for base is *alkali* because many, but not all, bases are ionic compounds of alkali metals or alkaline earth metals.

**Salts** are ionic compounds produced by the reaction of an acid with a base. Common salts are alkali or alkaline earth metals with halogens or polyatomic ions. Sodium chloride is table salt. Calcium chloride is a salt used in some ice-removal products. Although compounds of elements of groups 1 and 2 with group 17 are common salts, any metal ion can form a salt if it was part of the acid or base involved in the reaction. Positive polyatomic ions can form salts.

It took more than two hundred years for scientists to understand the behavior of acids and bases and to refine the definitions of *acid, base,* and *salt* to include all substances that share the same behaviors.

## Characteristics of Acids and Bases

In the 1600s, Robert Boyle first grouped substances as acids and bases according to characteristics such as taste, touch, and chemical reactions.

- Acids are sour like lemons. Bases are bitter like baking soda.

- Acids burn to the touch like vinegar. Bases are slippery like soap.

- Acids turn litmus (a dye made extracted from lichen) red. Bases turn litmus blue.

Figure 8–1. Litmus colors

- Acids react with metals to release hydrogen gas. Bases do not react with metals.

Acid solution     Metal solid     Hydrogen gas     Salt solution

$$2\,HCl_{(aq)} \quad + \quad Zn_{(s)} \quad \rightarrow \quad H_{s\,(g)} \quad + \quad ZnCl_{2(aq)}$$

Figure 8–2. Reaction equation metal with acid

- Acids react with carbonates to release carbon dioxide gas. Bases do not react with carbonates.

Acid solution    Carbonate compound    Carbon dioxide gas    Salt solution    Water liquid

$$2\,HCl_{(aq)} \quad + \quad NaHCO_3 \quad \rightarrow \quad CO_{s\,(g)} \quad + \quad NaCl_{2(aq)} \quad + \quad H_2O_{(l)}$$

Figure 8–3. Reaction equation for carbonates with acid

## Solutions of Acids and Bases

By the late 1800s, Swedish scientist Svante Arrhenius defined acids as substances that release hydrogen ions ($H^+$) when dissolved in water and bases as substances that release hydroxide ions ($OH^-$) when dissolved in water. When an acid and a base are combined, they produce water and a salt.

**Acid Solution in Water**

$$HCl \quad \xrightarrow{H_2O} \quad H^+_{(aq)} \quad + \quad Cl^-_{(aq)}$$

**Base Solution in Water**

$$NaOH \quad \xrightarrow{H_2O} \quad Na^+_{(aq)} \quad + \quad OH^-_{(aq)}$$

**Reaction of Acid and Base**

Acid     Base     Water     Salt

$$HCl \quad + \quad NaOH \quad \rightarrow \quad H_sO_{(aq)} \quad + \quad NaCl_{(aq)}$$

Figure 8–4. Equations for acid-base reactions

The Arrhenius definitions worked for most bases, but did not explain how substances that did not have a hydroxide group could behave as a base in reactions with acids. For example, carbonate compounds react with acids to form water, salt, and carbon dioxide gas.

# Defining Acids and Bases

As human knowledge of atomic structure increased, the definitions of acids and bases changed to reflect a greater understanding of the chemical interactions taking place in the formation of solutions.

Working independently, Danish chemist Johannes Bronsted and English chemist Thomas Lowry reached the conclusion that **acids** are substances that release hydrogen ions (proton) in water solutions and **bases** are substances that accept hydrogen ions in water solution. Acids and bases are opposites that react to form ionic compounds called salts. This definition explains why both hydroxide and carbonate compounds act as bases.

More complicated theories of acids and bases will be studied in advanced chemistry courses. For now, focus on properties and patterns of the reactions.

# Strong or Weak

Both acids and bases **dissociate** or break up into ions in water solution. **Strong acids** and **strong bases** easily become ions in solution. **Weak acids** and **weak bases** do not release very many ions in solution.

The terms *weak* and *strong* refer to how well the substances conduct electricity in solution. Because strong acids and bases release many ions, they are strong conductors of electricity. Because weak acids and bases release few ions, they are weak conductors of electricity.

Strong acids and strong bases are ionic compounds because the bonds break easily in water solution. Weak acids and weak bases are molecular compounds because the bonds are more difficult to break in water solution.

Hydrogen normally forms covalent bonds and behaves as a molecule. When hydrogen bonds with chloride, bromide, iodide sulfate, nitrate, or chlorate ions, it behaves like a metal and easily forms ions. This is shown in the results of a conductivity test of water, hydrochloric acid, and acetic (ethanoic) acid.

Strong bases are ionic compounds of the Group 1 and Group 2 metal element with the polyatomic ion hydroxide.

| The 7 Strong Acids $H^+$ in solution | The 8 Strong Bases $OH^-$ in solution |
|---|---|
| HI hydroiodic acid | NaOH sodium hydroxide |
| HBr hydrobromic acid | KOH potassium hydroxide |
| $HClO_4$ perchloric acid | LiOH lithium hydroxide |
| HCl hydrochloric acid | RbOH rubidium hydroxide |
| $HClO_3$ chloric acid | CsOH cesium hydroxide |
| $H_2SO_4$ sulfuric acid | $Ca(OH)_2$ calcium hydroxide |
| $HNO_3$ nitric acid | $Ba(OH)_2$ barium hydroxide |
| | $Sr(OH)_2$ strontium hydroxide |

Figure 8–5. Table of strong acids and bases

**Caution—Major Mistake Territory!**

*Strong* and *weak* **do not mean** "concentrated" or "dilute."

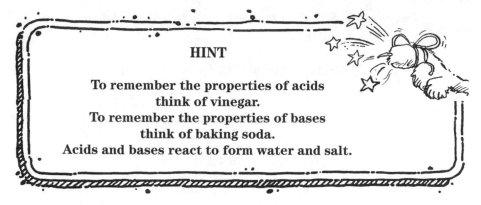

**HINT**

To remember the properties of acids
think of vinegar.
To remember the properties of bases
think of baking soda.
Acids and bases react to form water and salt.

# BRAIN TICKLERS
## Set #37

1. A food substance has a bitter taste and turns red litmus paper blue. Is the substance an acid, base, or salt?

2. Limestone is a rock that contains calcium carbonate. A drop of acid on limestone will bubble and fizz. What gas is released by this reaction?

3. Zinc powder was added to a clear liquid. Heat was released along with tiny gas bubbles. Was the liquid an acid or a base?

4. Liquid A turns blue litmus paper red. Liquid B turns red litmus paper blue. When the two liquids were combined, the resulting liquid C did not change the color of red litmus paper and did not change the color of blue litmus paper. What was liquid A? What was liquid B? What was liquid C?

*Questions 5–8 refer to the compounds tested in solution with a conductivity tester.*

| Answer Letter | Compound formula | Lightbulb |
|---|---|---|
| A | $H_2SO_4$ | Bright |
| B | NaOH | Bright |
| C | KOH | Bright |
| D | $H_2CO_3$ | Dim |
| E | $NaHCO_3$ | Dim |

5. Identify the compounds in the table as an acid or a base.

6. Identify the compounds in the table as weak or strong based on conductivity of electricity.

7. Which compounds behave as ions in solution?

8. Which compounds have strong bonds?

(Answers are on page 207.)

# pH SCALE

Sulfuric acid is one of the seven strong acids, meaning that it's a strong conductor of electricity in solution because it breaks up completely in water solution. This does not mean that it will be strongly acidic if it is in a dilute solution. In other words, one drop of sulfuric acid in a bathtub of water will likely not be noticed. The strong acid would be weakly acidic when diluted.

A better description of an acid or base is **acidity** or **alkalinity** measured as the concentration of hydrogen in solution. Acidity and alkalinity are expressed as the power of hydrogen or **pH**.

The **pH scale** ranks substances from 0 to 14 based on the concentration of hydrogen ions in solution. Each integer on the scale represents a negative exponent of 10. A change in pH of one integer is a change by a power of 10.

At pH 4, the concentration of hydrogen ions [H⁺] is $1 \times 10^{-4}$ moles/liter. At pH 5, the concentration of hydrogen ions [H⁺] is $1 \times 10^{-5}$ moles/liter. A pH 4 is more acidic than a pH 5. The mathematics for explaining this is complicated. Focus on the concepts associated with the pH scale. (Molarity as moles/liter is described in Chapter 5.)

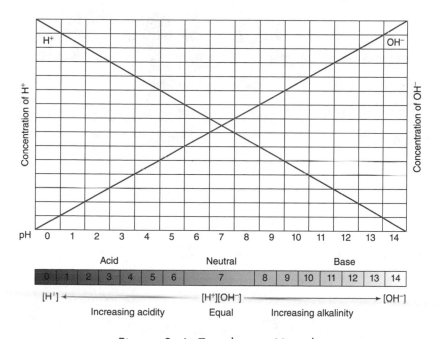

Figure 8–6. Trends on pH scale

Notice that as the pH decreases, the acidity increases because the concentration of hydrogen ions increases. Substances with a pH less than 7 are acids.

As the pH increases, the concentration of hydroxide ions increases. Substances with a pH greater than 7 are bases.

At pH 7, the concentration of hydrogen ions and hydroxide ions are in balance. The substance is neither acid nor base; it is neutral.

Because pH is an exponent of 10, a pH can be a decimal value. A substance with pH 5 is more acidic than a substance with pH 5.5. A substance of pH 8 is less alkaline than a substance of pH 8.5.

## Indicators

Indicators are substances that undergo a change in color as pH changes. The indicator does not react to form a product. The table shows the color changes of five commonly used indicators.

| Indicator | Color in Acid pH < 7 | Color in Neutral pH = 7 | Color in Base pH > 7 |
|---|---|---|---|
| Litmus paper | red | no change | blue |
| Bromothymol blue | yellow | green blue | blue |
| Phenolphthalein | clear | clear | purple |
| Universal indicator | red | yellow green | purple |
| Red cabbage juice | red | green | blue |

## pH of Common Substances

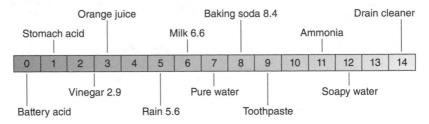
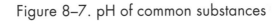

Figure 8–7. pH of common substances

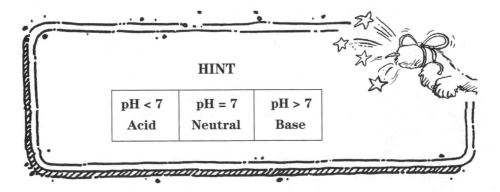

HINT

| pH < 7 | pH = 7 | pH > 7 |
|---|---|---|
| Acid | Neutral | Base |

# BRAIN TICKLERS
### Set #38

1. Universal indicator was added to a clear liquid. The liquid turned green. Which conclusion does the indicator evidence support?
   a. The liquid is an acid.
   b. The liquid has a pH greater than 7.
   c. The liquid is pure water.
   d. The liquid has a neutral pH.

2. Which statement correctly describes the relationship between a substance A that has a pH of 8 and a substance B that has a pH of 10?
   a. Substance A is a stronger base than Substance B because it has a lower pH.
   b. Substance B is more alkaline than Substance A because it has a higher pH.
   c. Substance A is more alkaline than Substance B because it has a lower pH.
   d. Substance B is a stronger base than Substance A because it has a higher pH.

3. When lemon juice is added to black tea, there is a color change. Select the best inference statement that explains why.

   a. Black tea contains chemicals that act as acid pH indicators.

   b. Lemon juice is a lighter color liquid that dilutes the black tea color.

   c. Black tea is a base and lemon juice is an acid.

   d. Lemon juice is a strong acid and black tea is a weak acid.

4. If normal rain is pH 5.6, what effect would you expect it to have on rocks containing calcium carbonate?

5. A mixture of water and universal indicator in a beaker was yellow green in color. Iron chloride was added. The color turned red. Analyze the reaction equation and explain the color change.

$$FeCl_3 + 3\ H_2O\ \rightarrow\ Fe(OH)_2 + 3\ HCl$$

$$HCl\ \rightarrow\ H^+ + Cl^-$$

(Answers are on pages 207–208.)

# NEUTRALIZATION REACTIONS

Neutralization reactions are a type of double replacement reaction that begins with an acid and a base as the reactants and ends with water and a salt as the products. Salts are ionic compounds. Not all salts have a neutral pH in solution.

When a strong acid and a strong base react, the salt product is usually neutral. The reaction of hydrochloric acid and sodium hydroxide yields sodium chloride and water. Sodium chloride is a neutral salt that is used as a seasoning in food. (Note: The molecular formula of water has been written as HOH to show the double replacement reaction.)

$$HCl + NaOH\ \rightarrow\ NaCl + HOH$$

Figure 8–8. Equation of neutral salt

When a strong acid reacts with a weak base, the salt product is an acidic salt. Sulfuric acid reacts with ammonium hydroxide to produce ammonium sulfate and water. Ammonium sulfate is an acidic salt that is used in fertilizers to lower the pH of alkaline soils.

$$H_2SO_4 + 2\ NH_4OH\ \rightarrow\ (NH_4)_2SO_4 + 2\ HOH$$

Figure 8–9. Equation of acid salt

An alkaline salt is the product of the reaction between a weak acid and a strong base. When carbonic acid reacts with sodium hydroxide, the products are water and the alkaline salt sodium carbonate. Sodium carbonate is also called washing soda and is used in the manufacturing of sodium bicarbonate or baking soda.

$$H_2CO_3 + 2\,NaOH \rightarrow Na_2CO_3 + 2\,H_2O$$

Figure 8–10. Equation of alkaline salt

The pH of a salt product of a reaction between a weak acid and a weak base depends on which reactant is relatively weaker. Think of a tug-of-war. The product pH will move toward the stronger reactant.

The pH of a salt product in solution will be important in more advanced chemistry courses. For now, visualize the concept that ions in a salt pull apart in solution and that loose ions can affect solution pH.

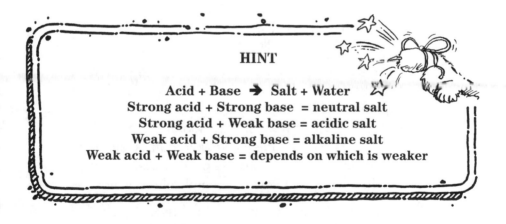

**HINT**

Acid + Base ➔ Salt + Water
Strong acid + Strong base = neutral salt
Strong acid + Weak base = acidic salt
Weak acid + Strong base = alkaline salt
Weak acid + Weak base = depends on which is weaker

## BRAIN TICKLERS
### Set #39

*Questions 1 and 2 refer to the reaction of acetic acid and sodium bicarbonate shown in the following equation.*

$$HC_2H_3O_{2(aq)} + NaHCO_{3(aq)} \rightarrow$$
$$NaC_2H_3O_{2(aq)} + HOH_{(l)} + CO_{2(g)}$$

1. Acetic acid (vinegar) is a weak acid. Sodium bicarbonate (baking soda) is a weak base. Which of the following would demonstrate the pH of the salt?
   a. metric scale
   b. thermometer
   c. indicator
   d. flame test

2. After the carbon dioxide bubbled off, the sodium acetate salt solution pH was 8. Which of the following explains why?
   a. Acetic acid is weaker than sodium bicarbonate.
   b. Sodium bicarbonate is weaker than acetic acid.
   c. Carbon dioxide raises the pH of water.
   d. The acetic acid was not completely reacted.

3. A strong acid reacts with a weak base. Which statement is true?
   a. The salt will have a pH of 7.
   b. The salt will have a pH less than 7.
   c. The salt will have a pH greater than 7.
   d. The neutrality, acidity, or alkalinity of the salt cannot be predicted.

4. Which of the following compounds is **NOT** a salt?
   a. $KNO_3$
   b. $H_2SO_4$
   c. $Na_2CO_3$
   d. $NH_2C_2H_3O_2$

(Answers are on page 208.)

## SUMMARY OF CHARACTERISTICS OF ACIDS AND BASES

| Characteristic | Acid | Base |
|---|---|---|
| Taste | Sour | Bitter |
| Touch | Burns | Slippery |
| Reacts with litmus paper | Red | Blue |
| pH | Less than 7 | Greater than 7 |
| Reaction of acid and base | Produces salt and water pH raises to 7 | Produces salt and water pH lowers to 7 |
| Reaction with metal | Releases hydrogen | No reaction |
| Reaction with carbonates | Releases carbon dioxide | No reaction |
| In water solution | Releases $H^+$ | Accepts $H^+$ |
| Compounds contain | $H^+$ | $OH^-$ or $HCO_3^-$ |
| Conductivity of electricity in solution | Strong if ionic Weak if molecular | Strong if ionic Weak if molecular |

## Wrapping Up

- Acids release hydrogen ions in solution.

- Strong acids break up in water more easily than weak acids.

- Ionic acids are strong acids and molecular acids are weak acids.

- Bases accept hydrogen ions in solution.

- Strong bases break up in water more easily than weak bases.

- Ionic acids are strong bases and molecular acids are weak acids.

- Acids and bases react to form water and salt.

- Salts are ionic compounds formed by neutralization reactions of acid and bases.

- The power of hydrogen, pH, is a measure of acidity.

- Acids are substances with a pH less than 7.

- Neutral substances have a pH of 7.

- Bases are substances with a pH greater than 7.

- Another word for base is *alkali*. The pH range above 7 is referred to as alkalinity.

- An indicator can be used to show if a substance is acid, neutral, or base.

- The pH of a salt depends on the strength of the acid and the strength of the base that react to form the salt.

# BRAIN TICKLERS—THE ANSWERS

## Set #37, pages 197–198

1. base

2. carbon dioxide

3. acid (hydrogen bubbles are tiny)

4. Liquid A is an acid. Liquid B is a base. Liquid C was salt water.

5. Acids are A and D. They will release $H^+$ ion in solution. Bases are B, C, and E. They are compounds of a metal and hydroxide or carbonate.

6. A is a strong acid. B and C are strong bases. D is a strong acid. E is a weak base.

7. A, B, and C will behave as ions in solution.

8. D and E have strong bonds and will behave as molecules in solution.

## Set #38, pages 201–202

1. d. The liquid has a neutral pH. (There are substances other than water that have a neutral pH.)

2. b. Substance B is more alkaline than Substance A because it has a higher pH. Nothing is known about the tendency of either substance to break up in solution. The only conclusion that can be made from the information is one of relative alkalinity.

3. a. Black tea contains chemicals that act as acid pH indicators. None of the other answers is supported by the observation.

4. Because normal rain has an acid pH, it will react with calcium carbonate in rocks to release carbon dioxide. (This process is called chemical weathering.)

5. The reaction started at a neutral pH. The iron chloride reacted with water to form iron hydroxide, a base, and hydrochloric acid. The color change indicates that the products are acidic. This means that hydrochloric acid is a strong acid that releases hydrogen ions in solution. Iron hydroxide is probably a weak base that does not easily break apart to release hydroxide ions in solution.

## Set #39, pages 204–205

1. c. indicator

2. a. Acetic acid is weaker than sodium bicarbonate.

3. b. The salt will have a pH less than 7.

4. b. $H_2SO_4$

# CHAPTER NINE

# Nuclear Reactions

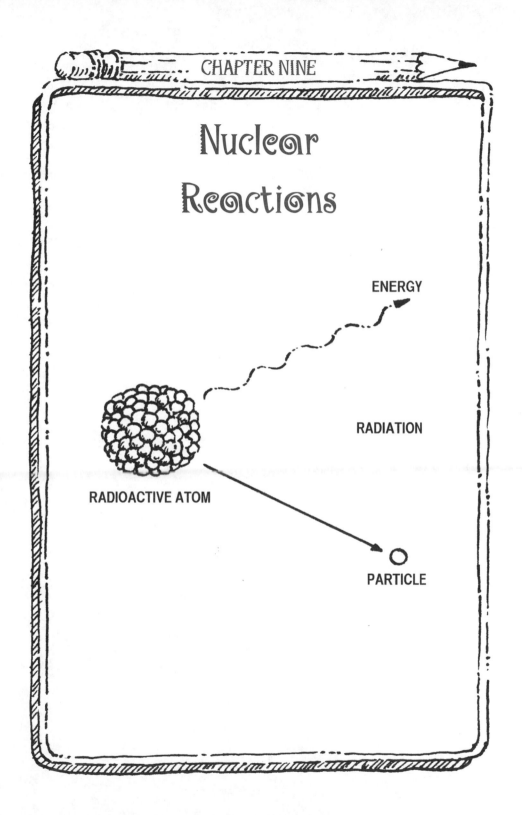

ENERGY

RADIATION

RADIOACTIVE ATOM

PARTICLE

# CHEMICAL REACTIONS VERSUS NUCLEAR REACTIONS

Chemical reactions involve the valence electrons of atoms. During chemical reactions, atoms rearrange to form new substances, but the elements of the reactants are the same elements of the products.

**Nuclear reactions** involve the core of the atom and form new elements. The three main nuclear reactions are nuclear fission, nuclear fusion, and radioactive decay.

# NUCLEAR FISSION

**Nuclear fission** reactions split the nucleus of an atom. The resulting atoms are elements with a lower atomic number than the original atom. Nuclear fission releases energy.

Nuclear fission occurs in certain isotopes of heavy elements. Uranium-235 is the most common fissionable isotope. Other isotopes that are used in atom-splitting reactions are plutonium-239, thorium-232, and uranium-233.

## Discovery of Nuclear Fission

During the 1930s, German physicists Otto Hahn and Fritz Strassman set out to create a heavier element from uranium-235 by striking the nucleus with a neutron. They were surprised to discover that their attempt generated lighter elements rather than heavier elements.

Other scientists forced to escape from Europe during World War II joined scientists in the United States as part of the Manhattan Project. The Manhattan Project research produced the atom bombs that were dropped on Hiroshima and Nagasaki, Japan, in 1945. The bombs demonstrated the enormous power of uncontrolled nuclear fission reactions.

# Uranium-235 Fission Reaction

The diagram shows a typical fission reaction. A low-energy neutron is absorbed into the nucleus of the uranium-235 isotope atom. This causes the nucleus to become unstable, breaking up into the elements krypton and barium and releasing three neutrons.

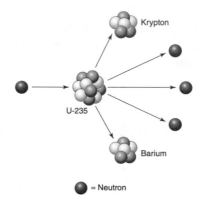

Figure 9–1. Uranium fission

The neutrons are free to continue striking U-235 atoms. The chain reaction caused by the free neutrons releases heat. Nuclear reactors use this heat to convert liquid water to steam to turn turbines that produce electricity.

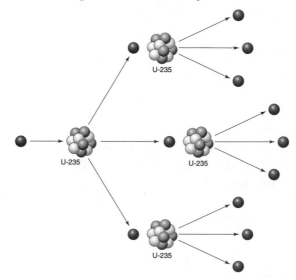

Figure 9–2. Uranium fission chain reaction

Nuclear power is an alternative to fossil fuel combustion because no greenhouse gas is produced in the reaction. However, there are other environmental concerns. Nuclear waste is radioactive and requires special disposal procedures. There is the potential for release of radioactive materials, and an uncontrolled nuclear reaction could be catastrophic.

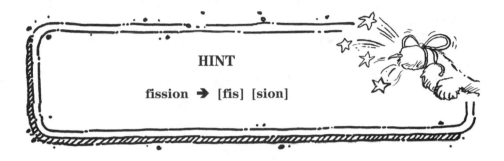

HINT

fission ➜ [fis] [sion]

## BRAIN TICKLERS
### Set #40

*Decide if the statement is true or false. Rewrite the false statements as true statements.*

1. Nuclear fission reactions produce heavier elements.

2. Electrons are used to start a nuclear fission chain reaction.

3. Heat is released by nuclear fission reactions.

4. Nuclear reactors emit greenhouse gases.

5. Uranium-238 is the unstable uranium isotope used in nuclear reactors.

(Answers are on page 225.)

# NUCLEAR FUSION

**Nuclear fusion** reactions combine the nuclei of two light atoms to form a new element. Nuclear fusion reactions naturally occur in the sun and other stars.

Nuclear fusion in stars generates heavier elements from lighter elements. For example the fusion of two hydrogen nuclei produces helium. The fusion of helium nuclei can produce carbon and oxygen. Our sun uses fusion to make elements as heavy as iron.

## Nuclear Fusion of Hydrogen

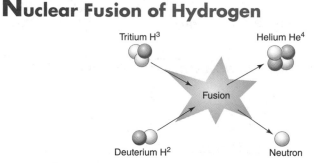

Figure 9–3. Nuclear fusion

The diagram shows the nuclear fusion of the hydrogen isotopes tritium $H^3$ and deuterium $H^2$. To achieve nuclear fusion, the tritium and deuterium gases are heated to about 100,000,000° Celsius. This converts the gases into plasma, pure nuclei stripped of electrons.

After the tritium and deuterium gases are converted to plasma, the nuclei fuse to form helium $He^4$, a free neutron, and energy. Notice that the mass of the new helium nucleus is less than the combined mass of the tritium and deuterium. This is accounted for by the mass of the free neutron.

## Harnessing the Power of Nuclear Fusion

On Earth, tritium is not a common isotope of hydrogen. It can be produced in large quantities by the nuclear fission of lithium-6, an abundant element in Earth's crust. When a neutron strikes the nucleus of lithium-6, the products are tritium and helium. Deuterium is a common isotope of hydrogen that can be extracted from water.

The materials to produce energy by nuclear fusion are readily available, and relatively small quantities produce tremendous amounts of energy. However, the high temperature needed to produce nuclear fusion reactions is a drawback for nuclear fusion power plants.

To date, controlled nuclear fusion reactions have been possible only on a small scale inside special reactors. An international research effort is under way to develop technology for controlling nuclear fusion reactions on the scale needed to produce safe, clean energy as an alternative to fossil fuels.

## Comparison of Fusion and Fission Reactions

- Nuclear fusion produces more energy than nuclear fission reactions.

- Unlike nuclear fission reactions, nuclear fusion reactions do not produce chain reactions. If the conditions needed for a nuclear fusion reaction break down, the reaction stops.

- Nuclear fusion reactions use the hydrogen isotopes tritium and deuterium. Nuclear fission reactions rely on unstable isotopes of heavy elements. However, tritium is radioactive and is a potential environmental hazard.

- Nuclear fusion reactions produce helium and neutrons. The products of nuclear fission are radioactive and pose an environmental disposal problem.

- Capture of neutrons produced by a nuclear fusion reaction pose an engineering challenge in the design of large-scale fusion reactors.

- Nuclear fusion requires extremely high temperatures compared with nuclear fission.

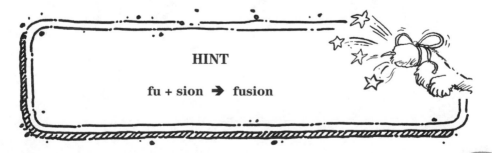

**HINT**

fu + sion ➜ fusion

# BRAIN TICKLERS
## Set #41

1. Describe the main difference between a nuclear fission reaction and a nuclear fusion reaction.

2. Why is the mass of the helium product less than the total mass of the tritium and deuterium reactants?

3. Where do natural nuclear fusion reactions occur?

4. How many helium nuclei would be needed to produce a carbon nucleus? (Hint: Use the periodic table of the elements.)

5. What phase of matter is needed for a nuclear fusion reaction?

6. Which of the following is a major drawback to the large-scale production of energy from nuclear fusion?
   a. The reactant materials are expensive.
   b. The reactant materials are rare.
   c. The reaction products are radioactive.
   d. The reaction requires extremely high temperatures.

(Answers are on page 226.)

# RADIOACTIVE DECAY

Nuclear fission and nuclear fusion require a collision between a neutron and a nucleus. **Radioactive decay** is the natural breakdown of an unstable nucleus. A nucleus will undergo decay to reduce the energy generated by an imbalance between protons and neutrons. Radioactive decay explains why atoms have isotopes.

The three main types of radioactive decay are alpha, beta, and gamma. The names are the first three letters of the Greek alphabet. Nuclei undergoing **alpha decay** release a helium nucleus that can be stopped by a sheet of paper. **Beta decay** releases either an electron or positron that can be stopped by a sheet of aluminum. **Gamma decay** releases high-energy gamma rays that are stopped by lead. Gamma decay often follows alpha or beta decay.

Alpha and beta decay create a new element atom from an existing element atom. Gamma decay stabilizes an atom. The atom that decays is called the **parent atom**. The atom that results from the decay is called the **daughter atom**.

## Alpha Decay

Alpha decay occurs in heavy elements. The nucleus of the parent atom releases an alpha particle. An alpha particle is two protons and two neutrons or the nucleus of a helium atom. Most of the helium in Earth's atmosphere comes from alpha decay of heavy elements in Earth's crust.

This loss of four nuclear particles reduces the atomic mass of the parent by four atomic mass units. The loss of two protons reduces the atomic number by two. The new daughter element atom is lighter than the parent element atom.

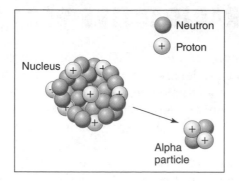

Figure 9–4. Alpha decay

## Example

$$^{238}_{92} \text{U} \quad \rightarrow \quad ^{234}_{90} \text{Th} \quad + \quad ^{4}_{2} \text{He}$$

In this example, uranium-238 emits a helium nucleus and becomes thorium-234. Notice that thorium is four atomic mass units lighter than uranium-238 and the atomic number of thorium is two less than uranium. Uranium is an element found in Earth's crust.

# Beta Decay

Beta decay emits a beta particle. Protons and neutrons are made of smaller particles called quarks. Energy changes in quarks can cause one type of quark to become another. This can happen in a proton or in a neutron. There are two types of beta decay: beta minus decay and beta plus decay.

During **beta minus decay**, a neutron becomes a proton. The neutron loses a particle equivalent to an electron ($e^-$) and another particle called an antineutrino ($v_e^-$). One of the down quarks in a neutron becomes an up quark. When that happens, the neutron converts to a proton.

The daughter atom has one more positive charge than the parent atom, but has the same atomic mass. The daughter atom is a new element.

## Example

$$^{14}_{6} \text{C} \quad \rightarrow \quad ^{14}_{7} \text{N} \quad + \quad e^- + v_e^-$$

The carbon-14 isotope decays into nitrogen-14. Carbon-14 is found in a ratio to carbon-12 that allows scientists to estimate the age of organic materials that are less than 40,000 years old.

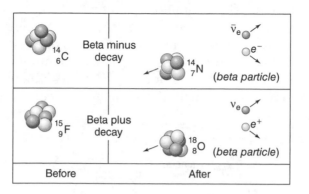

Figure 9–5. Beta decay

During **beta plus decay**, a proton becomes a neutron. Beta plus decay requires energy. The decay releases a positron ($e^+$) and a neutrino ($v_e$). The atomic number of the daughter atom is one less than the parent atom because it has one less proton. Because the proton becomes a neutron, the sum of the protons and neutrons does not change. Therefore, the mass of the daughter atom is the same as the mass of the parent atom.

## Example

$$^{18}_{9}F \quad \rightarrow \quad ^{18}_{8}O \quad + \quad e^+ + v_e$$

Water molecules made of the heavier oxygen-18 isotope condense at higher temperatures than water molecules made of oxygen-16 molecules. The water molecules with heavier oxygen will fall as rain in warmer latitudes. The rain in polar latitudes will be rich in oxygen-16 molecules. The ratio of oxygen-18 to oxygen-16 can be used to predict what climates were like in ancient times. High levels of oxygen-18 indicate a warmer climate, whereas high levels of oxygen-16 indicate a cooler climate. Ice cores from polar glaciers can reveal thousands of years of climate history.

## Gamma Decay

When the nucleus of an element has too much energy, it becomes unstable. This commonly happens after alpha or beta decay. Gamma decay reduces the energy of the nucleus. The nucleus releases gamma radiation in the form of a photon that is also known as a gamma particle. The atomic number and atomic mass of the atom do not change because the number of protons and neutrons remain the same. No new element forms.

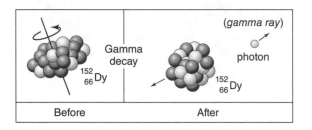

Figure 9–6. Gamma decay

### Example

$$^{152}_{96} \text{Dy} \quad \rightarrow \quad ^{152}_{96} \text{Dy} \quad + \quad \text{photon}$$

Gamma rays are part of the electromagnetic spectrum and have a very short wavelength with high frequency. X-rays are an artificial form of gamma radiation. Gamma rays are dangerous because of their ability to penetrate the cells of living things. They can force an atom to lose an electron, causing an atom in the cell to become a positive ion. The ionized atom can then react with other atoms, possibly damaging the cell or causing cancer, the uncontrolled growth of cells.

## Half-life

Elements that undergo natural radioactive decay usually do so in a predictable time frame. The term *half-life* is used to define how long it takes for half the atoms of an element to undergo decay. The time depends on the element. Some elements such as uranium-238 have a half-life of billions of years, and others such as radon-220 are measured in seconds. The environmental importance of this is discussed in Chapter 10.

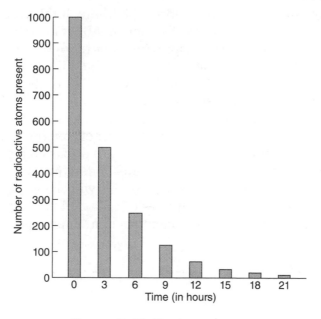

Figure 9–7. Gamma decay

Look carefully at the half-life graph. The starting sample is 1,000 atoms. After three hours, 500 atoms of the original element remain and 500 atoms have become something new. Three hours later, 250 atoms or half of the 500 atoms remain. The decay continues with half of the atoms left at the end of each three-hour period.

Because half of the atoms decay every three hours, the half-life is three hours. It is not possible to predict exactly which atom will undergo decay. It is possible to predict how much of an element will remain after a certain amount of time and how much of the decay products will be present after a certain amount of time.

Think of radioactive decay as popping corn. You know that most of the kernels will pop early in the cooking time, that popping will slow down toward the end, and that some kernels will remain unpopped at the end. You cannot at the beginning know exactly which kernels will pop first, in the middle, at the end, or not at all. However, based on experience, you can estimate how long it will take for half the kernels to pop, knowing that popping will slow down as cooking time continues and that there will be unpopped kernels at the end.

Now, think of popcorn in terms of half-life. It takes about 180 seconds to microwave a bag of popcorn. About half the kernels pop after 45 seconds. That leaves one-quarter of the kernels left to pop. Half of those kernels pop in the next 45 seconds. That leaves one-eighth of the kernels left to go. Half of those kernels pop in the next 45 seconds. In the last 45 seconds half of the remaining one-sixteenth of the kernels pop. One-thirty-second (half of one-sixteenth) of the original kernels did not pop.

Like the popcorn model, the majority of atoms change during the first half-life. The number of atoms undergoing decay slows down over time, but half of what remains will change during each half-life. After many half-lives of radioactive decay, some of the original atoms of an element remain "unpopped."

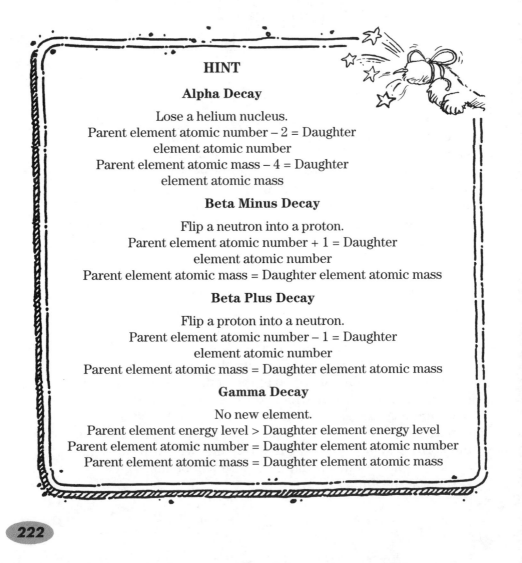

## HINT

### Alpha Decay

Lose a helium nucleus.
Parent element atomic number – 2 = Daughter element atomic number
Parent element atomic mass – 4 = Daughter element atomic mass

### Beta Minus Decay

Flip a neutron into a proton.
Parent element atomic number + 1 = Daughter element atomic number
Parent element atomic mass = Daughter element atomic mass

### Beta Plus Decay

Flip a proton into a neutron.
Parent element atomic number – 1 = Daughter element atomic number
Parent element atomic mass = Daughter element atomic mass

### Gamma Decay

No new element.
Parent element energy level > Daughter element energy level
Parent element atomic number = Daughter element atomic number
Parent element atomic mass = Daughter element atomic mass

# BRAIN TICKLERS
### Set #42

*Use the periodic table of the elements as needed.*

| Element Atomic Number | Thorium-232 Decay Series Half-life of Each Element | | | | |
|---|---|---|---|---|---|
| Thorium 90 | Th-232 $1.39 \times 10^{10}$ yrs. | | Th-220 1.90 yrs. | | |
| Actinium 89 | ↓ | Ac-228 ↗ 6.13 hrs. | ↓ | | |
| Radium 88 | Ra-228 ↗ 6.7 yrs. | | Ra-224 3.64 days | | |
| Francium 87 | | | ↓ | | |
| Radon 86 | | | Rn-220 54.5 sec. | | |
| Astatine 85 | | | ↓ | | |
| Polonium 84 | | | Po-216 0.16 sec. | 65% | Po-212 $3.0 \times 10^{-7}$ sec. |
| Bismuth 83 | | | ↓ | B-212 ↗ 60.5 min. | ↓ |
| Lead 82 | | | Pb-212 ↗ 10.6 hrs. | 35% ↓ | Pb-208 stable |
| Thallium 81 | | | | Th-208 ↗ 3.1 min. | |

1. What type of decay transforms thorium-232 into radium-228? Explain your answer.

2. What type of decay transforms radium-228 into actinium-228? Explain your answer.

3. Which element in the decay chain has the shortest half-life?

4. Why does the decay chain end in lead-208?

5. Which of the following emits an electromagnetic energy wave?
   a. alpha decay
   b. beta minus decay
   c. beta plus decay
   d. gamma decay

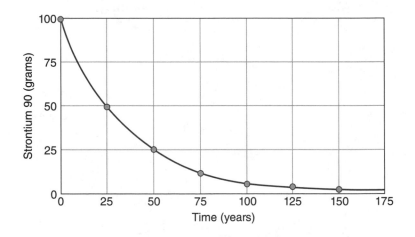

6. According to the graph, what is the half-life of strontium-90?

7. How many years will it take for the sample to contain only 6.25 grams of strontium-90?

(Answers are on page 226.)

## Wrapping Up

- Nuclear fission is the splitting of two atomic nuclei.
- Nuclear fission produces nuclear power to generate electricity.
- Nuclear fusion is the joining of two atomic nuclei.
- Nuclear fusion occurs naturally in stars.
- Radioactive decay is a natural process that stabilizes high-energy atomic nuclei.
- Alpha decay releases a helium nucleus. The atomic number of the parent atom decreases by two and the atomic mass decreases by four.
- Beta minus decay flips a neutron into a proton, releasing an electron and an antineutrino. The atomic number of the parent atom increases by one and the atomic mass is unchanged.
- Beta plus decay flips a proton into a neutron, releasing a positron and a neutrino. The atomic number of the parent atom decreases by one and the atomic mass is unchanged.
- Gamma decay releases a shortwave electromagnetic gamma radiation. There is no change.

# BRAIN TICKLERS—THE ANSWERS

## Set #40, page 213

1. False—Nuclear fission reactions produce lighter elements.

2. False—Neutrons are used to start a nuclear fission chain reaction.

3. True

4. False—Nuclear reactors do not emit greenhouse gases.

5. False—Uranium-235 is the unstable uranium isotope used in nuclear reactors.

## Set #41, page 216

1. Nuclei are divided by fission and joined by fusion.

2. A neutron is lost in the fusion reaction.

3. the sun and other stars.

4. Three helium nuclei produce one carbon-12 nucleus.

| helium | 2 protons, 2 neutrons |
|---|---|
| + | |
| helium | 2 protons, 2 neutrons |
| + | |
| helium | 2 protons, 2 neutrons |
| carbon-12 | 6 protons, 6 neutrons |

5. plasma

6. d. The reaction requires extremely high temperatures.

## Set #42, pages 223–224

1. Thorium-232 loses a total of 4 atomic units. The atomic number decreases by 2 protons. The decay of thorium-323 to radium-228 is alpha decay with the loss of a helium nucleus.

2. The atomic mass of radium-228 is the same as actinium-228. The atomic number of actinium-228 is one more than radium-228. A neutron must change to a proton. This is beta plus decay.

3. polonium-212

4. Lead-208 is stable, meaning that the energy in the nucleus is not too high or too low.

5. d. gamma decay

6. 25 years

7. 100 years (4 half-lives)

# CHAPTER TEN

# Environmental
# Chemistry

# VOLCANIC ERUPTIONS

Volcanic eruptions are natural events that cycle resources from Earth's lithosphere (rock) into the atmosphere (air) and hydrosphere (water). The chemicals released by a volcano are unique to the location of the volcano. Scientists can use the chemical signature to study how materials from a volcanic eruption are spread through the atmosphere.

Radioactive elements below Earth's surface undergo decay. As they break down, they release heat. Earth's crust (upper rock layer) acts like a pressure cooker, increasing the heat energy.

The extremely high temperatures melt solid rock into a thick liquid mixture called magma. Magma is less dense than solid rock; therefore it rises to the surface through cracks and vents. When magma reaches the surface, it erupts as lava.

At high temperatures, chemical compounds in rocks break down and undergo chemical reactions. As the magma reaches the surface, pressure decreases and hot dissolved gases expand. Lava explodes, releasing these gases, metal compounds, and glasslike particles of ash.

There are five main gases produced by volcanoes: water vapor, carbon dioxide, sulfur dioxide, fluorine, and chlorine. All of the gases with the exception of water vapor are harmful to living things.

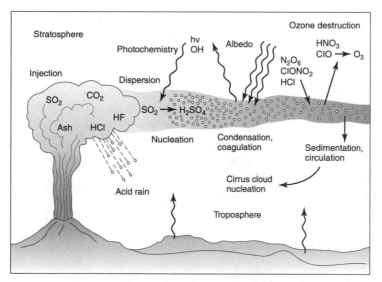

Figure 10–1. Volcanic gases

Volcanic eruptions are a source of carbon dioxide. Carbonates (carbon and oxygen compounds) lock carbon into certain types of rocks. When the rocks are cycled into Earth's mantle and melted, carbon dioxide is freed as a gas. Carbon dioxide traps heat from the sun, causing the greenhouse effect.

In Earth's past during periods of high volcanic activity, volcanoes may have had an effect on global climate. Volcanoes are not a significant cause of modern increases in global temperatures related to rising levels of atmospheric carbon dioxide.

Sulfur dioxide forms an aerosol that reflects sunlight. After major eruptions, sulfur dioxide aerosol can spread globally and cool temperatures for years. Sulfur dioxide also reacts with water vapor to form sulfuric acid that can fall as acid rain. In Earth's past, sulfur dioxide aerosols from frequent volcanic eruptions may have caused cool periods.

Fluorine is deadly when inhaled and harmful to both plants and animals. Rainwater washes fluorine out of the atmosphere onto plants. Animals that eat the contaminated plants experience fluorine poisoning.

Chlorine quickly reacts with water vapor to form hydrochloric acid. Chlorine reacts with the ozone in the stratosphere, disrupting the layer that protects the Earth from ultraviolet radiation from the sun. Ultraviolet radiation damages DNA and can affect the health of both plants and animals.

Although weathering and erosion are more important than volcanoes in metal cycling, volcanoes release iron, manganese, mercury, and other metals and metalloids. Metals such as iron are micronutrients that are needed by phytoplankton and bacteria for growth.

Volcanic ash resembles glass more than the light fluffy ash from a campfire. The high temperatures melt silicates that cool into sharp, jagged particles. The ash can cause severe damage to the respiratory systems of animals and death. Ash also interferes with jet engines of airplanes and etches the windows of the cockpit.

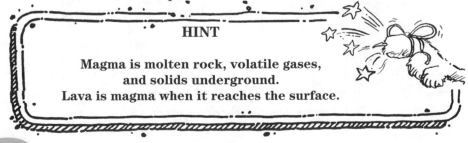

### HINT

Magma is molten rock, volatile gases, and solids underground.
Lava is magma when it reaches the surface.

## BRAIN TICKLERS
### Set #43

**Complete the statements.**

1. Sulfur dioxide aerosols reflect sunlight, causing a _____ effect.

2. Sulfur dioxide aerosols react with _____ to form sulfuric acid.

3. Carbon dioxide causes a _____ effect.

4. Under reduced pressure, volatile gases in magma _____ as they reach the surface.

5. The ozone layer is affected by _____ formed from chlorine gas.

6. Magma is an example of _____ (a mixture, an element, a compound)

(Answers are on page 252.)

# SMOG FORMATION

**Smog**, or smoky fog, is a dangerous mixture of particles, ground-level ozone, and other chemicals that can affect the health of both animals and plants. Smog forms when pollution from motor vehicles and factory smokestacks becomes trapped, allowing concentrations to increase. Energy from the sun is the catalyst for the reactions that lead to smog formation.

Smog is more common in cities that experience air inversions. Normally, warm air rises from ground level spreading pollutants out at higher altitudes. During an inversion, the air at ground level is cooler than the air above it. The dense air tends to remain at ground level. The warmer air above the cooler air acts as a lid on a pot of simmering water trapping the pollutants.

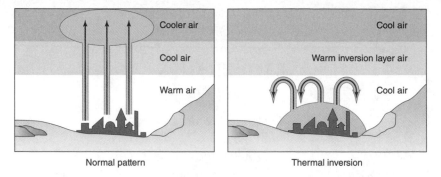

| Normal pattern | Thermal inversion |

Figure 10–2. Inversion

Inversions can occur because of horizontal winds from the ocean. In California, breezes from the Pacific Ocean blow from west to east. Air rising up mountains to the east cools, becomes dense and begins to fall back to the surface. As the air returns to the surface it warms becoming warmer than the air blowing from the west. The inversion layer traps pollutants.

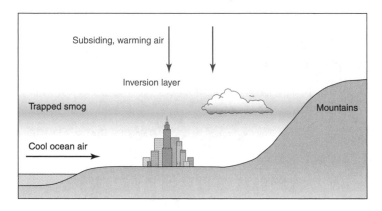

Figure 10–3. Advection inversion

Along the East Coast, land breezes carry pollutants east to the Atlantic Ocean. In late afternoon winds shift to sea breezes that push the pollution west and back onto land. The pollutants in the cooler air become trapped below warmer air.

In the Midwest, inversions occur in valley regions. At night cooler, dense air flows down the sides of hills and mountains into the valley. The cooler air is trapped under the warmer air it displaces. During an inversion the valley acts as a bowl with a lid, holding pollutants at ground level.

General conditions for smog formation are (1) high concentrations of nitrogen oxides and volatile organic compounds,

(2) weather with clear sky, direct sunshine, and calm winds, and (3) air temperatures above 90°F (32.2°C).

Smog formation is a series of chemical reactions. The main sources of reactants are cars, trucks, and buses. The tailpipe gases include nitrogen oxide, volatile organic compounds, and carbon monoxide.

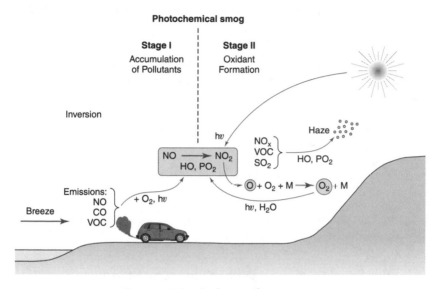

Figure 10–4. Smog formation

In the first step of smog formation, nitrogen oxide combines with oxygen to form nitrate gas. This gas quickly reacts with nitrogen oxide to form nitrogen dioxide, an irritating reddish-brown gas.

$$NO \quad + \quad O_2 \quad \rightarrow \quad NO_3$$

$$NO_3 \quad + \quad NO \quad \rightarrow \quad 2NO_2$$

Figure 10–5. Nitrogen dioxide equation

Nitrogen dioxide will slowly decompose into nitrogen oxide and a free oxygen ion. With sunlight as a catalyst, the oxygen ions quickly react with oxygen molecules in the air to form ground-level ozone, a gas that irritates the eyes and respiratory system.

$$NO_2 \quad \rightarrow \quad NO \quad + \quad O$$

$$O \quad + \quad O_2 \quad \rightarrow \quad O_3$$

Figure 10–5. Ozone formation equation

Volatile organic compounds are highly reactive hydrocarbons emitted from tailpipes and smokestacks. The compounds form aldehydes that as small aerosol particles can trigger asthma attacks.

Nitrogen dioxide, ground-level ozone, and the reaction products of volatile organic compounds cause the haze associated with smog. These chemicals are extremely dangerous to those who have respiratory conditions such as asthma or emphysema and can be irritating to healthy individuals. Air-quality laws are helping to prevent smog formation by reducing the concentration of the reactants.

# BRAIN TICKLERS
## Set #44

**Select the correct answer to each question.**

1. What is the catalyst in smog formation?
   a. nitrogen oxides
   b. water vapor
   c. sunlight
   d. volatile organic compounds

2. Which of the following happens during an inversion?
   a. Warmer air is trapped below cooler air.
   b. Cooler air is trapped below warmer air.
   c. A cool air, warm air, cool air sandwich forms.
   d. A warm air, cool air, warm air sandwich forms.

3. Where would smog formation occur with greatest frequency?
   a. rural areas
   b. suburban areas
   c. forested areas
   d. urban areas

4. What is the major source of reactants that form smog?
   a. tailpipes
   b. smokestacks
   c. lightning
   d. power plants

5. Which of the following is an irritating reddish-brown gas that forms during the smog reaction?
   a. ozone
   b. sulfur dioxide
   c. volatile organic compounds
   d. nitrogen dioxide

6. What is the molecular formula of ozone?
   a. $O_2$
   b. $O_3$
   c. $O_4$
   d. $O_2$–$O_2$

(Answers are on page 252.)

# ACID RAIN

All rain is acidic. In the atmosphere water vapor and carbon dioxide react to form weak carbonic acid with a pH of approximately 5.6. The acidic pH of natural rain has limited effect on life on Earth.

$$H_2O + CO_2 \rightarrow H_2CO_3$$

water + carbon dioxide $\rightarrow$ carbonic acid

Figure 10–6. Carbonic acid equation

Water in the atmosphere will also react with nitrogen oxides and sulfur dioxide, forming acids with a pH lower than 5.6. The dry particles of these oxides and the wet acids that form with water are together known as **acid rain**. Dry deposition describes particles and gases that coat surfaces at ground level. Wet deposition describes precipitation as acid rain, snow, or fog.

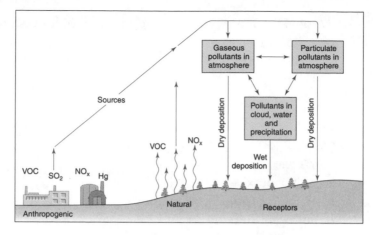

Figure 10–7. Sources of pollution

*Anthropogenic* means "caused by humans." Volcanoes emit sulfur dioxide and nitrogen oxides, but most of the pollutants that cause acid rain have human sources.

Sulfur dioxide comes primarily from the combustion of coal. Coal-burning power plants are the major sources of sulfur dioxide. When sulfur dioxide reacts with oxygen, it becomes sulfur trioxide. The sulfur trioxide reacts with water to make sulfuric acid.

$$H_2O + SO_3 \rightarrow H_2SO_4$$

water + sulfur trioxide → sulfuric acid

Figure 10–8. Sulfuric acid equation

Tailpipes and factory smokestacks are the major sources of nitrogen oxides. Lightning is a minor natural source of nitrogen oxides in the atmosphere. The reaction of nitrogen dioxide with water produces nitric acid and nitrogen oxide. Nitrogen oxide can react with oxygen to make more nitrogen dioxide.

$$3NO_2 + H_2O \rightarrow 2HNO_3 + NO$$

nitrogen dioxide + water → nitric acid + nitrogen oxide

Figure 10–9. Equation for nitric acid formation

Acid rain is a problem because acids interact with carbonates, eating away at marble and limestone monuments, buildings, rock formations, and soil.

$$CaCO_3 + H_2SO_4 \rightarrow CaSO_4 + H_2O + CO_2$$
calcium carbonate + sulfuric acid ➜ calcium sulfate + water
+ carbon dioxide

Figure 10–10. Equation for calcium carbonate and sulfuric acid

Acid rain also reacts with metals in soils. Reduction in soil quality affects plants, killing trees and crops. Pollutants carried by wind currents can travel great distances. Pollution from the Midwest has affected forests and lakes in the Adirondack Mountains of New York.

In aquatic ecosystems, acid rain disrupts shell formation by mollusks (soft-bodied animals with calcium carbonate shells). Rainbow trout begin to die at pH 6. At pH 5.5, frog eggs, tadpoles, crayfish, and mayflies die. When lake and stream pH drops to 4.2, all fish die.

Air-quality laws have reduced the human sources of nitrogen oxides and sulfur dioxide. By limiting the reactants, the acid rain products are reduced. Areas that were strongly affected by acid rain are showing signs of recovery.

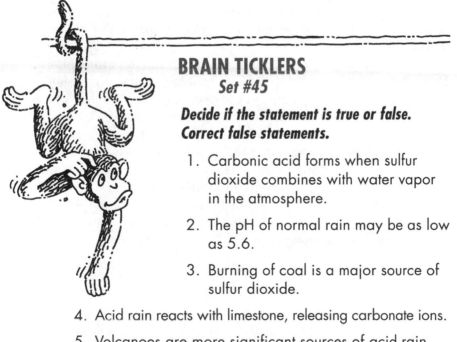

## BRAIN TICKLERS
### Set #45

**Decide if the statement is true or false. Correct false statements.**

1. Carbonic acid forms when sulfur dioxide combines with water vapor in the atmosphere.

2. The pH of normal rain may be as low as 5.6.

3. Burning of coal is a major source of sulfur dioxide.

4. Acid rain reacts with limestone, releasing carbonate ions.

5. Volcanoes are more significant sources of acid rain reactants than human activities.

(Answers are on pages 252–253.)

# OCEAN ACIDIFICATION

Earth has one ocean because all oceans and seas are connected. Names are assigned to geographic regions of large expanses of salt water bordered by land. Whether water comes from an ocean or a sea, it is referred to as seawater.

Seawater is a mixture of water, minerals, metals, and salts. When mixed in water, the salts form ions. The average pH range of sea surface water is 7.5 to 8.4. Seawater is alkaline (a base).

Minerals, metals, and salts enter ocean waters by runoff from land or by settling of dust from air over the ocean. Iron, manganese, and other metals form hydroxide compounds in seawater. Nitrates and phosphates from fertilizers dissolve in seawater, becoming available nutrients for phytoplankton, the producers of the ocean.

Gases enter the ocean from the atmosphere. If a gas is in higher concentration in the air above the ocean water, gas molecules will dissolve into the ocean until a balance is reached. If a gas is in higher concentration in the ocean waters than in the air above the ocean, the gas will diffuse out of the water into the air.

Nitrogen gas is the most abundant gas in Earth's atmosphere. Dissolved nitrogen is converted into usable nitrogen compounds by nitrogen-fixing organisms in the ocean.

The cycling of carbon in the ocean is complicated. Carbon dioxide is used by phytoplankton for photosynthesis. Carbon dioxide is waste produced by the consumer organisms that eat phytoplankton or other consumers. Decomposing organisms release carbon dioxide. Carbon dioxide is trapped and removed from the ocean through limestone (calcium carbonate rock) formation and calcium carbonate skeletal formation by animals such as clams and corals.

Ocean Acidification

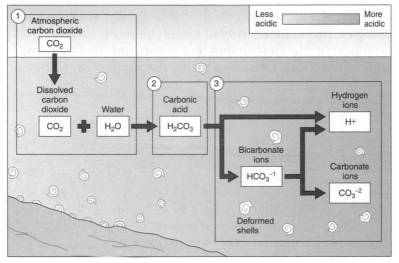

Figure 10–11. Ocean acidification

**Ocean acidification** occurs when the carbon dioxide entering the ocean from the atmosphere overwhelms the natural carbon cycling of ocean systems. Carbon dioxide diffuses into the ocean, forming weak carbonic acid. The carbonic acid breaks up into bicarbonate ions and hydrogen ions. Bicarbonate ions decompose into carbonate ions and more hydrogen ions. Hydrogen ions lower the pH of the ocean.

The formation of carbonic acid can be observed by placing bromothymol blue in a glass of water. When carbon dioxide is added to the water, the solution turns from blue green (neutral pH) to green yellow (acid pH). If the solution is allowed to stand for a few hours, the water will turn blue-green as excess carbon dioxide leaves the water and reenters the atmosphere.

The ocean will not reach an acid pH (less than 7) because it is a mixture of salts and weak bases. When carbon dioxide is bubbled into a solution of seawater, the color of the bromothymol blue remains blue. After a high concentration of carbon dioxide has been bubbled into the salt water, the salt water will turn blue green (neutral), but will not become yellow (acid).

Although the ocean is unlikely to become an acid bath, the lower pH caused by increased hydrogen ions affects the ability of living things to survive in the ocean habitat. The lower pH interferes with the ability of mollusks to form calcium carbonate shells and limits coral reef building by blocking calcification of the coral skeleton. The lower pH also affects the ability of some zooplankton to make outer shells. A global research effort is under way to understand the chemistry of the interaction of the atmosphere and ocean and the impact of chemical reactions occurring below the surface of the ocean.

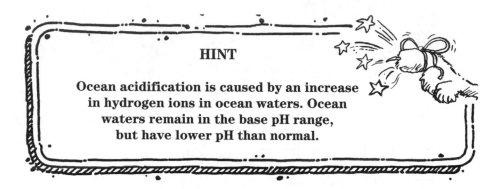

### HINT

Ocean acidification is caused by an increase in hydrogen ions in ocean waters. Ocean waters remain in the base pH range, but have lower pH than normal.

# BRAIN TICKLERS
## Set #46

### Select the correct answer.

1. What forms first when carbon dioxide dissolves in seawater?
   a. calcium carbonate
   b. carbonic acid
   c. hydrogen chloride
   d. bicarbonate ions

2. What will happen to the concentration of dissolved carbon dioxide in seawater if carbon dioxide concentration in the atmosphere increases?
   a. The concentration of carbon dioxide in seawater will increase so that there is more carbon dioxide in oceans than in the atmosphere.
   b. The concentration of carbon dioxide in seawater will decrease as carbon dioxide returns to the atmosphere.
   c. The concentration of carbon dioxide in seawater will be unchanged because carbon dioxide will be trapped in coral reef skeletons.
   d. The concentration of carbon dioxide in seawater will increase until it is in balance with carbon dioxide in the atmosphere.

3. Why is ocean acidification a threat to organisms that depend on calcium carbonate for skeleton formation?
   a. An increased concentration of hydrogen ions decreases pH, changing the balance between calcium carbonate, bicarbonate ions, and carbonate ions.
   b. An increased concentration of hydroxide ions reduces the pH of the ocean, causing more bicarbonate ions to form than carbonate ions.
   c. An increased concentration of hydrogen ions increases pH, changing the balance between calcium carbonate, bicarbonate ions, and carbonate ions.
   d. An increased concentration of hydroxide ions increases the pH of the ocean, causing more bicarbonate ions to form than carbonate ions.

4. Bromothymol blue was added to a sample of seawater. The indicator remained blue. What does this mean?
   a. The sample has a neutral pH.
   b. The sample has an acidic pH.
   c. The sample has an alkaline (base) pH.
   d. No conclusion about pH can be reached.

(Answers are on page 253.)

# ATMOSPHERIC OZONE

Earth's atmosphere is composed of four main layers: the troposphere, the stratosphere, the mesosphere, and the thermosphere. These layers of gases and ions protect the surface of the Earth from harmful radiation from the sun.

The ozone layer is found in the stratosphere at an altitude of about 5.5–7.5 miles (9–12 km) from Earth's surface. Visible light and some ultraviolet radiation pass through the ozone layer.

At ground level in the troposphere, ozone is a dangerous air pollutant that can cause respiratory distress and other harm to animals and plants. In the stratosphere, ozone blocks or reflects most of the harmful ultraviolet A (UVA) and ultraviolet B (UVB) wavelengths so that they never reach the troposphere. Ultraviolet radiation causes sunburn and skin cancer, and can harm plants and animals other than humans.

A molecule of oxygen is composed of two oxygen atoms that equally share two pairs of electrons in a double bond. Ozone is composed of three oxygen atoms. Electrons are partially shared by all three oxygen atoms. This creates areas of positive and negative charges or two poles (dipole). The valence electrons move around all three oxygen nuclei.

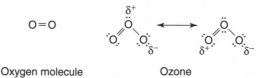

Oxygen molecule                    Ozone

Figure 10–12. Ozone and oxygen molecule

Natural processes in the stratosphere produce atmospheric ozone. Ultraviolet radiation from the sun strikes an oxygen molecule, splitting the molecule into two oxygen ions. The free oxygen atoms collide with other oxygen atoms, forming ozone.

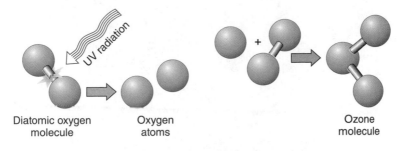

Figure 10–13. Ozone formation

Ozone is naturally broken down as it absorbs ultraviolet radiation. Ultraviolet radiation splits an oxygen atom from an ozone molecule, forming a molecule of oxygen and a free oxygen atom. The oxygen atom collides with another ozone molecule, causing it to break down into two oxygen molecules.

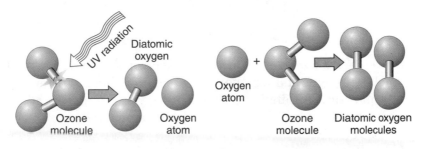

Figure10–14. Natural ozone depletion

The cycle of natural ozone formation and depletion by absorption of ultraviolet radiation shields the surface of the Earth from 95–99% of the harmful ultraviolet waves.

Chlorofluorohydrocarbons (CFCs) from human activities have disrupted the natural cycle of atmospheric ozone formation and destruction. Before their use was banned by most nations, compounds of chlorine, fluorine, and carbon known as CFCs were used in a number of consumer products and industrial processes as aerosol propellants and refrigerants. The hairspray can and refrigerator became symbols of CFC destruction of the ozone layer.

Molecules of CFCs are carried along Earth's circulation belts, collecting at the equator. Hot air containing CFCs rises from the troposphere and enters the ozone layer in the equatorial region. The CFC molecules spread out in the ozone layer, reaching the polar regions.

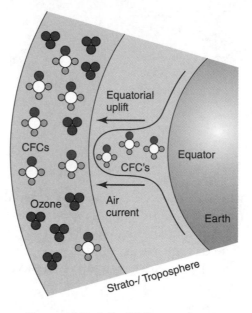

Figure 10–15. CFC circulation

Evidence for damage to Earth's ozone layer was first observed in satellite images and weather balloon data collected primarily over Antarctica. A thinning of the ozone layer, popularly called the hole in the ozone layer, occurs on a seasonal basis over Antarctica.

Laboratory experiments confirmed that CFCs could break down atmospheric ozone. When ultraviolet radiation strikes a CFC molecule, a chlorine atom breaks loose.

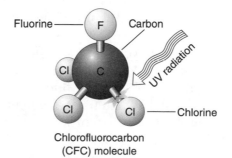

Chlorofluorocarbon
(CFC) molecule

Figure 10–16. CFC molecule

The free chlorine atom collides with an ozone molecule, forming chlorine monoxide and a molecule of oxygen.

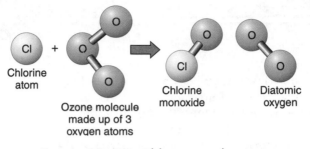

Figure 10-17. Chlorine and ozone

Oxygen molecules are broken up into free oxygen atoms by ultraviolet radiation. When a free oxygen atom strikes a chlorine monoxide molecule, the oxygen atom replaces the chlorine atom, producing a molecule of oxygen and a free chlorine atom. The chlorine atom is free to react with another ozone molecule.

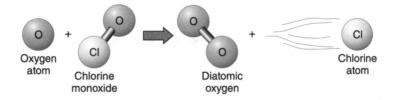
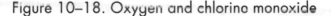

Figure 10-18. Oxygen and chlorine monoxide

Ultraviolet radiation is necessary to break a chlorine atom free from a CFC molecule. Once a chlorine atom is freed, the chain reaction can continue in the absence of ultraviolet radiation. Although the addition of new CFC into the atmosphere has been restricted, the chlorine continues to affect the natural ozone formation-depletion cycle.

Formation of new ozone requires ultraviolet radiation. This is why ozone-layer depletion continues during the Antarctic winter even though the continent is in darkness for months. The ozone layer begins repair once sunlight returns during the Antarctic spring.

Scientists continue to study the ozone layer and to monitor changes in atmospheric chemistry and level of ultraviolet radiation reaching Earth's surface.

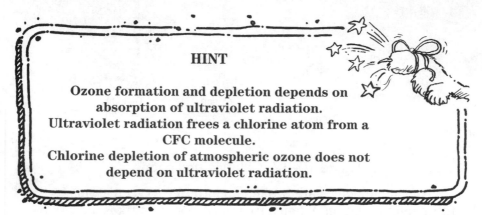

**HINT**

Ozone formation and depletion depends on
absorption of ultraviolet radiation.
Ultraviolet radiation frees a chlorine atom from a
CFC molecule.
Chlorine depletion of atmospheric ozone does not
depend on ultraviolet radiation.

# BRAIN TICKLERS
## Set #47

1. Where is the ozone layer located?

2. What do atmospheric ozone molecules absorb?

3. Why is ozone-layer depletion continuing even though CFC releases have been reduced?

4. How does the lack of sunlight during the winter affect the ozone layer over Antarctica?

(Answers are on page 253.)

# RADON-222 ALERT

Radon-222 is a chemically inert, radioactive decay product of uranium-238. Uranium-238 is a naturally occurring isotope found in igneous (magma) rocks such as granite. Uranium-238 decay daughters are more likely to be found in ancient continental bedrock.

Unlike other elements in the uranium-238 decay chain, radon-222 is a gas. Radon-222 collects in soil, rock layers, and water sources. Radon-222 gas seeps into homes, schools, and other buildings through the basement or foundation. Gas concentrations increase when they are trapped inside buildings as compared with outdoors, where air circulation allows molecules to spread out.

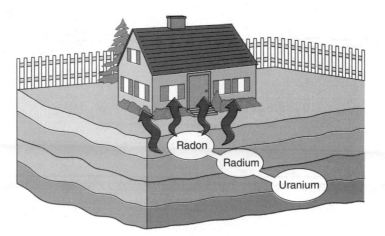

Figure 10–19. Radon gas pockets

Radon-222 can be unknowingly inhaled, because it is a colorless, odorless, tasteless gas. Radon-222 has a half-life of 3.382 days. When it decays, radon-222 releases a dangerous alpha particle.

The decay daughters of radon-222 include isotopes of polonium, lead, and bismuth that all have relatively short half-lives that are measured in minutes. Unlike radon-222, the decay daughters are solids that are chemically reactive as well as radioactive. When the decay daughters are suspended in air, they become attached to dust. These radioactive particles can then be inhaled, increasing exposure to alpha particles.

Radioactive decay of radon-222 is measured in picocuries per liter (pCi/L). A picocurie per liter is equivalent to 2.2 particle decays per minute in a 1-liter sample of a gas.

Average indoor levels of radon-222 are considered to be 1.3 pCi/L to 4 pCi/L. Although no level of radon-222 is completely safe, 4 pCi/L is considered the upper limit for safe exposure in home and school environments. This means that continuous exposure to levels above 4 pCi/L is considered hazardous. Radon-222 is the second leading cause of lung cancer in the United States.

Maps of radon-222 pockets are available from state environmental protection departments and the U.S. Environmental Protection Agency. Inexpensive home test kits are commercially available and recommended for homes in high-risk areas.

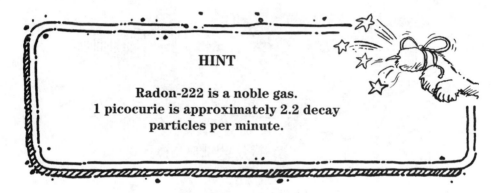

**HINT**

**Radon-222 is a noble gas.
1 picocurie is approximately 2.2 decay
particles per minute.**

# BRAIN TICKLERS
## Set #48

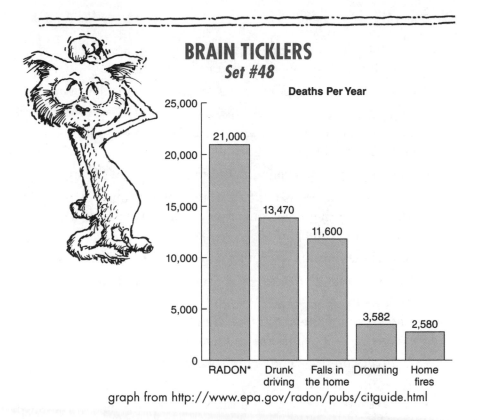

**Deaths Per Year**

graph from http://www.epa.gov/radon/pubs/citguide.html

1. According to the graph what is the estimated death rate from lung cancer related to radon-222 exposure?

2. Why is radon-222 more likely to cause lung cancer than other uranium-238 decay radioisotopes?

3. How many decay particles would be released at 4 pCi/L radon-222 levels?

4. What type of rock is radon-222 associated with?

5. Why doesn't radon-222 react with other elements to form compounds?

(Answers are on pages 253–254.)

## Wrapping Up

- Volcanoes have a unique signature that can be used to identify particles released during an eruption.

- Gases in magma expand as magma reaches the surface, leading to explosive eruptions.

- Volcanoes release water vapor, carbon dioxide, sulfur dioxide, fluorine, and chlorine.

- Volcanic eruptions can affect climate.

- Smog is a dangerous mixture of particles, ground-level ozone, and other chemicals that can affect the health of both animals and plants.

- Sunlight is the catalyst for ground-level ozone formation.

- The reactants that form smog are volatile organic compounds and nitrogen oxides.

- Smog is more likely to form when temperatures are in excess of 90°F, winds are calm, and skies are clear.

- An inversion traps cooler, dense air below warmer, less dense air.

- Carbon dioxide reacts with rainwater to form weak carbonic acid.

- Acid rain has a pH lower than 5.6.

- Sulfur dioxide reacts with rainwater to form sulfuric acid.

- The lower pH of acid rain affects the ability of plants and animals to survive in terrestrial and aquatic environments.

- Combustion of coal by power plants is a primary source of sulfur dioxide.

- Seawater acts as a buffer because it is a mixture of water, minerals, metals, and salts.

- Carbon dioxide diffuses into seawater until the concentration in the water is equal to the concentration in the air above the water.

- Carbon is cycled in the ocean through organic and inorganic chemical reactions.

- Carbon dioxide in seawater forms carbonic acid that breaks down in bicarbonate, carbonate, and hydrogen ions.

- Hydrogen ions lower the natural pH of seawater.

- Although seawater will remain in the alkaline (base) pH range, reduction in pH by excess hydrogen ions is called ocean acidification.

- A reduction in seawater pH negatively affects the ability of organisms to build calcium carbonate skeletons.

- The Earth's ozone layer is in the stratosphere approximately 5.5–7.5 miles (9–12 km) above Earth's surface.

- An ozone molecule is composed of three oxygen atoms.

- The natural cycle of atmospheric ozone formation and depletion depends on ultraviolet radiation from the sun.

- The absorption of ultraviolet A and ultraviolet B wavelengths by the ozone cycle protects life on Earth from excessive exposure to harmful radiation.

- Chlorofluorohydrocarbons (CFCs) disrupt the formation of ozone and increase ozone depletion.

- Free chlorine atoms released by absorption of ultraviolet radiation by CFCs can remain in the stratosphere for an underdetermined period of time.

- Once free chlorine atoms are released from CFCs, the depletion of atmospheric ozone does not depend on solar radiation.

- Radon-222 is a chemically inert radioactive gas that is part of the decay chain of uranium-238.

- Radon-222 is a colorless, tasteless, odorless gas that has a half-life of 3.382 days.

- Alpha radiation from radon-222 and the short-lived decay daughters of radon-222 are a leading cause of lung cancer.

- Radon-222 is associated with ancient igneous continental bedrock formations such as granite.

- Concentrations of radon-222 in excess of 4 picocuries per liter are considered hazardous.

# BRAIN TICKLERS—THE ANSWERS

## Set #43, page 231

1. cooling

2. water vapor

3. warming/heating

4. expand

5. hydrochloric acid

6. a mixture

## Set #44, page 234–235

1. c.  sunlight

2. b.  Cooler air is trapped below warmer air.

3. d.  urban areas

4. a.  tailpipes

5. d.  nitrogen dioxide

6. b.  $O_3$

## Set #45, page 237

1. False

   Carbonic acid forms when carbon dioxide combines with water vapor in the atmosphere.

2. True

3. True

4. False

   Acid rain reacts with limestone, releasing carbon dioxide gas.

5. False

   Human activities are more significant sources of acid rain reactants than volcanoes.

## Set #46, pages 240–241

1. b. carbonic acid

2. d. The concentration of carbon dioxide in seawater will increase until it is in balance with carbon dioxide in the atmosphere.

3. a. An increased concentration of hydrogen ions decreases pH, changing the balance between calcium carbonate, bicarbonate ions, and carbonate ions.

4. c. The sample has an alkaline (base) pH.

## Set #47, page 246

1. The ozone layer is in the stratosphere 5.5–7.5 miles (9–12 km) above the surface of the Earth.

2. Atmospheric ozone molecules absorb ultraviolet A and ultraviolet B radiation.

3. Ultraviolet radiation releases chlorine atoms from CFCs. The chlorine atoms remain in the stratosphere, reacting with oxygen and ozone.

4. Free chlorine atoms break ozone molecules down. Without ultraviolet radiation, ozone molecules are not formed as fast as chlorine breaks down ozone molecules.

## Set #48, page 249

1. The estimated death rate from lung cancer caused by radon-222 is 21,000 per year.

2. Radon-222 is a gas that can be inhaled and the half-life of radon-222 daughter elements is short.

3. 4 pCi/L × 22 decay particles/min/pCi/L = 8 to 9 decay particles/minute

4. Radon-222 would probably accumulate in areas with ancient granite formations.

5. Radon-222 is a noble gas. Each atom of radon-222 has a full valence; therefore it does not need to share electrons with other atoms.